COMMON CORE MATH in ACTION

Making the Standards Manageable, Meaningful & Fun

Catherine Jones Kuhns
Marrie Lasater

Crystal Springs
SDE BOOKS

a division of Staff Development for Educators

Peterborough, New Hampshire

Published by Crystal Springs Books
A division of Staff Development for Educators (SDE)
10 Sharon Road, PO Box 500
Peterborough, NH 03458
1-800-321-0401
www.SDE.com/crystalsprings

Printed in the United States of America
19 18 17 16   3 4 5

ISBN: 978-1-935502-76-0
e-book ISBN: 978-1-935502-77-7

Library of Congress Cataloging-in-Publication Data

Kuhns, Catherine Jones, 1952- author.
  Common core math in action : making the standards manageable,
    meaningful & fun, grades 3-5 / Catherine Jones Kuhns & Marrie Lasater.
    pages cm
  Includes bibliographical references and index.
  ISBN 978-1-935502-76-0 — ISBN 978-1-935502-77-7 (e-book) 1.
Mathematics--Study and teaching (Middle school)—Standards—United
States. 2. Middle school education—Curricula—Standards—United
States. 3. Mathematics—Study and teaching (Middle school—Activity
programs—United States. I. Lasater, Marrie, 1951- author. II. Title.

  QA13.K838 2015
  372.7—dc23

                        2014004028

To the many devoted teachers in our noble profession whom I have worked with from sea to shining sea (and beyond). —C.J.K.

To my family—you're the loves of my life—Judd, John, Sarah, and the soon-to-arrive 'Baby Lasater.'—M.L.

Contents

Acknowledgments 1

How to Use This Book 2

    Using the Assessments 5

Introduction 6

    History of the Common Core State Standards 6

    The Eight Standards for Mathematical Practice 6

    Concrete-Pictorial-Abstract 7

    Managing Tools & Other Resources. 9

    Taking Time for Questions & Answers 9

    Making It Meaningful 11

    Talking & Writing Mathematically 12

Domain: Operations and Algebraic Thinking. **15**

    Grade 3 16

    Grade 4 32

    Grade 5 42

Domain: Number and Operations in Base Ten **49**

    Grade 3 50

    Grade 4 55

    Grade 5 67

Domain: Number and Operations—Fractions **81**

    Grade 3 82

    Grade 4 95

    Grade 5 117

Domain: Measurement and Data **139**

    Grade 3 140

    Grade 4 164

    Grade 5 182

Domain: Geometry **199**

    Grade 3 200

    Grade 4 204

    Grade 5 210

Copymasters & Assessments 219

References & Resources 231

    Professional Books/Resources 232

    Children's Literature with Math Connections for Grades 3–5 233

    Websites 235

Index 236

Acknowledgments

Many thanks to the faculty and staff at Country Hills Elementary—where "what's right for kids" always comes first and foremost. I am honored to share the joy of teaching and learning with you! Deep gratitude also goes to Diane Lyons, editor extraordinaire. —C.J.K.

I would like to thank Teresa Jones, Lisa Jones, Christa Campbell, and the entire 4–5 team at McFadden School for allowing me to teach and "play" in your classrooms. The students in your classes make me proud to be a teacher. Thanks also to Dr. Clark Blair. Your leadership style combines strength and great love. And thank you to my dear friend, Cindy Cliche, for your continued support and guidance. I'm grateful to you, Cathy, for the opportunity to write with a veteran. —M.L.

How to Use This Book

The activities in this book are organized into five sections, one for each of the domains identified by the Common Core State Standards in Mathematics (CCSSM). Within each domain section, activities are organized by grade level, and within each grade level, by content standard.

Why are the activities grouped by domain, when the Common Core standards are grouped by grade level? Within each domain, there's significant overlap in concepts and skills across the grade levels. We decided to make it easy for you to glance through all of the activities for each domain, so you can quickly locate the best activities for your students' needs. Plus, it's helpful to revisit activities to keep skills solid from year to year.

Five Domains
Operations and Algebraic Thinking
Number and Operations in Base Ten
Number and Operations—Fractions
Measurement and Data
Geometry

At the start of each domain, we set the stage for developing a deep understanding of the concepts in that domain.

DOMAIN

Number and Operations in Base Ten

Here's the plain and simple truth about this domain: It requires procedural knowledge and rote learning. At first glance this might seem almost contradictory to what we think of as the core of CCSSM—critical thinking, communicating, precision in language, and reasoning. But take a closer look at each standard.

Yes, students are expected to develop fluency (the ability to solve a problem quickly and accurately), but check out these verbs used in each NBT standard—*recognize, illustrate, compare,* and *explain.* Students are also required to understand the procedures and strategies used in solving problems. They should not just be mindlessly repeating steps.

When students say and use the steps required to solve a problem efficiently, they should be doing so because they've internalized the reason for each step. When kids understand the procedures they're using, and can articulate why they're using them, you can rest assured that they'll be able to apply their knowledge to the more complex problems that are on their mathematical horizons.

Once again, we suggest that you look closely not only at the standards for your grade level, but also at the ones before and after the grade you teach. Third graders who learn to add and subtract numbers to 1,000 and master basic multiplication facts to 100 will transform into fifth graders who multiply and divide multidigit whole numbers and divide decimals. The leaps are huge!

We've deliberately selected many concrete and pictorial activities for this domain. Students would be well served if most of these activities were experienced more than once.

Whenever you can, seize the opportunities to connect your lessons to place value!

49

Domain sections contain activities for Grade 3, Grade 4, and Grade 5. Sections are color-coded to make it easy to see which grade you're in.

Activities are organized by cluster and by content standard, within each grade level.

C-P-A stages—concrete, pictorial, and abstract—and group size suggestions are identified for each activity.

CCSSM content standards are identified for every activity. The first activity that addresses a standard includes the full text of the standard; if more than 1 activity is provided, the later activities carry only the standard code.

The Standards for Mathematical Practice are critical to sound mathematical instruction. While each math activity touches on most, if not all, of the 8 mathematical practices, the practices addressed most prominently in each activity are identified (using a shorthand form).

DOMAIN
OPERATIONS AND ALGEBRAIC THINKING

③

C-P-A

Whole Group, Individuals

3.OA.A.1 Interpret products of whole numbers, e.g., interpret 5 × 7 as the total number of objects in 5 groups of 7 objects each. *For example, describe a context in which a total number of objects can be expressed as 5 × 7.*

Math Practices
7 Make Use of Structure
8 Express Regularity in Repeated Reasoning

3

GRADE ③

Cluster 3.OA.A Represent and solve problems involving multiplication and division.

Multiplication Masterpieces

Paul Klee (Getting to Know the World's Greatest Artists) by Mike Venezia and

Delicious: The Art and Life of Wayne Thiebaud by Susan Goldman Rubin

Examining a masterpiece or two of art each day not only sparks some culture and adds color to your math lesson, it also serves as a daily review of this standard.

To prepare, locate examples of works by Paul Klee, Wayne Thiebaud, or Wassily Kandinsky to share with your students. These artists all painted objects in arrays (Klee painted rectangles, Thiebaud food and other items, Kandinsky circles). Students will need tempera or watercolors, brushes and paper.

- Begin by displaying one piece of art at a time. When you show your students the first piece, ask them if they notice anything mathematical about the painting.

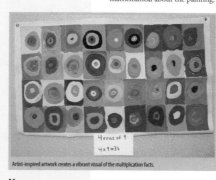

4 rows of 9
4 × 9 = 36

Artist-inspired artwork creates a vibrant visual of the multiplication facts.

- Discuss how the painting shows images in rows and columns. Explain that the total number of images can be calculated by multiplying the number of rows by the number of images in each row. (For example, a work by Thiebaud may show food items in 5 rows of 7, illustrating 5 × 7 = 35.) Examine at least one to two works of art at the beginning of each math class for a few days.

16

CCSSM CLUSTER CODES

The standards in the original print and pdf editions of <u>CCSS Mathematics</u> contain codes for grade level, domain, and standard; for example, "4.OA.2" is Grade 4, Operations and Algebraic Thinking, Standard 2. Following the publication of <u>CCSS for Mathematics</u>, the CCSSM writing team has added letter codes to identify the clusters at each grade level, in order to better facilitate communication and correlation. This book uses the system that includes cluster codes. In the new system, the same example standard becomes "4.OA.A.2"—Grade 4, Operations and Algebraic Thinking, Cluster A, Standard 2.

We're big advocates of math journals. Writing in a journal helps the writer to make sense of what he's learned and to keep track of mathematical thinking. In each domain there are **Write About It** activities that suggest ways your students can record their experiences in their math journals. You can encourage more math writing simply by saying, "Pull out those math journals and record what we just did." We've provided many examples of authentic well-written responses that demonstrate the child's grasp of the concept.

Many of our activities use children's literature as a springboard for mathematics. We know the power that a well-written and beautifully illustrated story holds for a child. If we can connect what we teach and say in one academic area to another, it's a win-win situation for our kids.

Write About It:
Have the students pictorially represent each layer of the stack. Children love to color their representations and show the abstract fraction that names each layer.

Stacking and drawing the pattern blocks help children visualize fractional equivalency.

Fraction Fun

Fraction Fun by David A. Adler

Adler's *Fraction Fun* is a delightful way to compare fractions and their relative sizes. The dog in this book is presented with 3 plates of pizza. Each plate has a different-sized slice. Will the dog choose $\frac{1}{2}$, $\frac{1}{4}$, or $\frac{1}{8}$? This story, along with the simple paper plate manipulatives, will get your students comparing fractions with ease. Start the activity by reading the first 13 pages of the book.

You'll need those paper plate fraction pieces you created in Paper Plate Fractions, page 88. These kid-friendly manipulatives will give your students the conceptual power to make a generalization. Generalizations and conjectures are what mathematical practice 8 is all about!

- Pass out the bags with the manipulatives. Tell students, "Take out 1 of each color except for the yellow one. Turn and talk to your neighbor about the size of each of the pieces. (Students should have $\frac{1}{2}$, $\frac{1}{4}$, and $\frac{1}{8}$.)
- Continue, "Let's pretend that the paper plate fractions are slices of pizza! Which piece would you want if you were very hungry?" Show the students the picture of the dog looking at the pieces of pizza in the story. "This dog is facing the same dilemma. Which

Whole Group

3.NF.A.3d Compare two fractions with the same numerator or the same denominator by reasoning about their size. Recognize that comparisons are valid only when the two fractions refer to the same whole. Record the results of comparisons with the symbols >, =, or <, and justify the conclusions, e.g., by using a visual fraction model.

Math Practices
4 Model with Mathematics
6 Attend to Precision
7 Make Use of Structure
8 Express Regularity in Repeated Reasoning

93

It takes time and repetition to develop deep understanding. We've suggested **Variations** to keep activities interesting when you revisit them, or just to try if you like the variation better than the original. Of course there are those students who get it right away and beg for more. Keep them busy and deepen their understanding with **Extensions**.

- Say, "All right, my Einsteins, here's the task for today. You'll notice the solid figures we used before. This time they're already filled with cubes. You'll also see some irregular solid figures that I built." Hold up a 1-inch cube and say, "This cube is 1 inch wide, 1 inch high and 1 inch deep. It is called a 1-inch cubic unit." (Or, if you're using metric figures, hold up a 1-centimeter cube and give the cube's metric dimensions.)
- Continue by saying, "Each solid figure is filled with 1-inch (or 1-centimeter) cubes. Your task is to determine the volume of each solid figure. You may need to reach inside to find out how many cubes high the shape is; in other words, you may need to find out how many layers there are since you can see only the top layer. It's that simple!"
- Say, "You'll each keep a record of your work in your math journal. For example, if this shape contained sixteen 1-inch cubes, then you'd write 'Shape B has a volume of 16 cubic inches.' We'll compare everyone's results once we're all done."
- First assign each small group a figure to begin with, and then allow children to move from one figure to the next.
- Challenge your students. Say, "While you're doing this activity, I want you to be thinking: Is there an easier, more efficient way to find volume other than counting all of the cubes?" (You're planting a seed here. Moving to the formula for finding volume is what the next standard is all about.)

Variation: After kids have gained experience with teacher-created figures, ask each group to create an irregular solid figure. Have the groups exchange their creations and determine the volume of their new figure. Let the original designers of each figure confirm the correct answer.

Extension: You can use sugar cubes for this activity, but if you do, the children will need to understand that it takes 4 sugar cubes to equal one 1-inch cube (the typical sugar cube is a $\frac{1}{2}$-inch cubic unit). This makes a great center for children who need more challenges.

Stacked sugar cubes are sweet way to show volume.

191

4

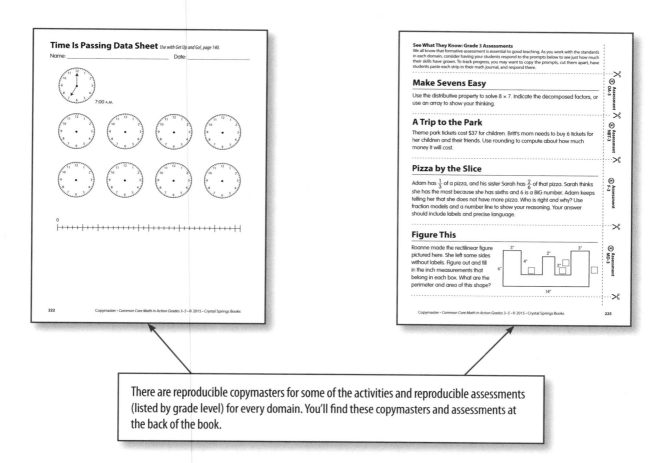

There are reproducible copymasters for some of the activities and reproducible assessments (listed by grade level) for every domain. You'll find these copymasters and assessments at the back of the book.

Using the Assessments

The assessments were designed for your students to showcase their understanding of the concepts developed, the vocabulary learned, and the mathematical thinking presented in each domain. You'll want to read over the problem with your class to be sure everyone understands the task. Once you're sure everyone grasps the problem, turn them loose! Tell students you want them to think and to have fun with the task.

As your students problem-solve, you need to circulate, listen, and offer encouragement. Remember, the first mathematical practice includes learning how to persevere! Take time to note what tools and strategies your students are using. Ask kids who finish early if they can figure out another way to solve the problem. Prompt struggling students with hints such as, "Can you solve this with any of the tools we've used?" or "Can you draw a picture or use a table or chart?"

Last, call the class together for a "Math Meeting" and invite students to share their strategies for solving the problem. Without pencils in hand (you don't want kids to change their answers), students should explain how they know their answers are right—the explanations are almost more important than the right answer!

Introduction

Welcome to our book! All of the activities here support specific Common Core State Standards for Mathematics (CCSSM) for grades 3–5. In fact, activities to support each and every one of the grade 3–5 standards are included! These activities were designed to engage your students in higher-level thinking and help them develop rich problem-solving strategies.

We're believers in the proverb, "Tell me and I'll forget; show me and I may remember; involve me and I'll understand." Our focus is on helping students develop a deep conceptual understanding of the mathematical concepts presented in the standards. We hope that you and your students will enjoy these math activities as much as we enjoy doing them with our mathematicians.

History of the Common Core State Standards

In June 2009, the Common Core State Standards Initiative (CCSSI) was undertaken by the National Governors Association Center for Best Practices and the Council of Chief State School Officers. Representatives from 48 states, 2 territories, and the District of Columbia, plus content experts, teachers, and researchers collaborated over the course of a year to draft the standards. Existing rigorous state standards and international standards were researched to further inform the development of the CCSSI.

Additionally, an advisory group including representatives from Achieve, Inc., ACT, the College Board, the National Association of the State Boards of Education, and the State Higher Education Executive Officers provided guidance. Thousands of public comments were taken into consideration during the drafting and revising of the standards documents before they were announced in their final form in June 2010.

The Eight Standards for Mathematical Practice

As you've discovered in your own experience with implementing the CCSSM, an integral part of the standards is the shift away from drill, rote formulas, and procedure (3 cheers!) to meaningful mathematics in which children make sense of problems, practice critical-thinking skills, and develop a deep understanding of concepts.

This emphasis is evident in the importance placed on the Standards for Mathematical Practice. While each grade level has its own specific content standards for each of the 5 domains, all grade levels share the same 8 practices across the domains. By applying these practices, we teachers can help our students develop into patient, competent problem-solvers and critical thinkers. While good math lessons and tasks utilize many of the 8 practices, it's often true that a few of the practices are most directly relevant. Those are the ones we've listed with each activity.

Concrete–Pictorial–Abstract

An important aspect of the CCSSM is a new emphasis on student engagement in mathematics. It isn't enough to get the right answer; children need to understand and explain why an answer is correct. In order to do this, children need a deep understanding of the mathematics—not just knowing their numbers and manipulating symbols. This kind of deep understanding starts with C-P-A.

This is not a new federal agency or a type of vitamin. It's simply a way to remember the stages of reasoning: concrete, pictorial, and abstract.

Using manipulatives is part of the *concrete* stage. Manipulatives are the best way to introduce a concept.

After plenty of practice with concrete materials, the *pictorial* stage helps bridge the gap between concrete and abstract. Seeing pictures and drawing pictures are both good ways for children to solidify understanding of a problem or concept. The authors of Common Core are careful to point out that student drawings need not show realistic details; they just need to illustrate the mathematics.

Last to develop is *abstract* understanding—the use of numerals and symbols to represent what's happening mathematically. It's important not to rush students into the abstract stage before they're ready. If we do, the mathematical meaning may be lost, and scaffolding of more complex concepts may be destined for disaster!

C ▸ P ▸ A An icon near each activity in this book denotes whether the activity develops mainly concrete, pictorial, or abstract reasoning. Some activities and their variations may be used to develop more than one of these stages; when this is the case, plan to revisit the activity over a period of time in order for your students to reach each level.

It will come as no surprise that many concepts in this book are introduced at the concrete or pictorial level. Our goal is to engage our students with interesting activities that promote great thinking. In order to develop deep understanding, children need to be exposed to concepts from more than a printed page.

C-P-A & COMMON CORE

Common Core is relatively new, but it's built on what we all learned in our early education classes from trailblazers like Maria Montessori, Jean Piaget, Zoltan Dienes, and Jerome Bruner: children learn best when they engage with real objects while solving real problems. The authors of Common Core also looked closely at the practices in other countries, such as Singapore, which consistently ranks at the top or near the top of all countries participating in the TIMSS (Trends in International Mathematics and Science Study). An emphasis on the concrete-to-pictorial-to-abstract progression is evident in every one of the countries with strong mathematics scores.

Managing Tools & Other Resources

Compasses, rulers, measuring tapes, number lines, protractors, angle rulers, cubes, geometric solids, pattern blocks, tiles, and place-value materials are all "tools" that should be made easily accessible to your students.

Math Practice 5, "Use appropriate tools strategically," means that students self-select proper tools to solve problems. With your guidance, children will learn how and when to use math tools, but self-selection can happen only if the materials are where kids can get their hands on them. Your praise for that self-initiative will go a long way when you want your students to take accountability for their learning.

The Dynamic Paper tool, a feature of the NCTM Illuminations website, is referenced frequently throughout this book. This tool gives teachers the ability to make customized grids, graphs, geometric shapes, and number lines with just a few clicks on the computer. The Dynamic Paper tool is absolutely free and it can be found at illuminations.nctm.org.

Taking Time for Questions & Answers

Developing a deep understanding of mathematics doesn't happen in a day, a week, or even a month. It takes time—lots of time—for introducing new concepts, thinking about them, talking about them, practicing them, and revisiting them. Children need time to

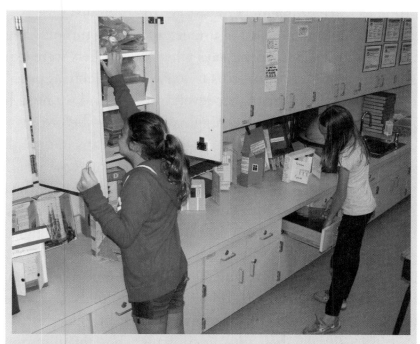

We like to keep our tools within easy reach so that students have the freedom to select the tool that works best for them!

Ask good questions and wait for thoughtful answers. It takes time, but it's worth it!

discuss their mathematical discoveries, to make mistakes, and to correct each other—in short, they need time to solve problems, not just get the right answers.

We can help students achieve the kind of understanding Common Core requires by asking thoughtful questions. Gone are the days when we teachers were expected to rattle off as many questions as possible for students to answer as quickly as possible. We need to ask fewer questions, ones that require higher-level thinking, and we need to give children time to think about the questions—even struggle with them—before answering.

Marian Small suggests that teachers ask open-ended questions in order to allow any student to find something appropriate to contribute to the conversation. Our questions should leave room for a variety of strategies to be used (Small, 2012). The question, "Is the number 7 prime or composite?" for example, is not open-ended. It has a single, correct, one-word answer, and the child has a 50/50 chance of getting it right simply by guessing.

By learning to rephrase the questions we ask, we can prompt more thoughtful responses. For example, if we instead ask, "How are the numbers 2, 3, 7, and 11 alike?" or "How are the numbers 3 and 7 different from the numbers 4, 8, and 9?" we invite students of differing abilities to take their time to construct thoughtful responses. Questions of this type take careful planning on our part, but the results are much more revealing than *yes, no,* or another one-word answer (Smith, 2013).

Have you ever opened your mouth and answered a question right away that you wish you had waited 60 seconds to answer? We all have. Would your answer have been better constructed if you had waited? Probably. Well, the same is true for our students. It's important to give children time to think and reflect. Refining an answer is a skill, and that skill takes time to practice and develop.

We can almost hear you wondering, "How can I find time to slow down?" Common Core carefully scaffolds the content we're teaching our students. By addressing fewer skills at each grade level, students have more time to go deeper and achieve mastery and we teachers have more time to construct questions that require more thinking.

Making It Meaningful

Good math instruction does more than just cover content. Research has shown that for a math lesson to be successful, for students to understand and retain it, the lesson should not just involve important mathematical concepts. It should begin with a problem that's interesting, and it should connect to prior knowledge (Hiebert et al., 2000).

The goal of excellent mathematics instruction is to create and lead engaging lessons in which the students enjoy using mathematics, deepen their understanding, connect their learning to the outside world, and learn new skills. One of our favorite education quotes,

Engaging, hands-on math lessons makes learning more meaningful for students.

simple yet profound, is from Joseph Renzulli, professor of educational psychology at the University of Connecticut and a researcher in the area of giftedness. Speaking to an auditorium filled with teachers, he said, "When the activity is engaging and entertaining, there will be more achievement." The activities in this book serve as a way for students to reach that level of joy, understanding, connection, and mastery.

We've developed deep and engaging activities for each standard. We admit, however, that your students will need several experiences with the same concept to truly "get" that concept. Many of the standards are huge, and engaging in just 1 or 2 activities with that concept will just hit the tip of that iceberg. Work with and share ideas with fellow teachers so that you can build up an arsenal of strong concept-building tasks. Think beyond the worksheet for solid understanding.

Some of the concepts within the standards may come much more slowly to your kiddos than you had hoped. It could well be that your students didn't come to you with the necessary background knowledge. In these cases, refer to materials for the grade level below the one you teach. The progression of the standards was very carefully thought out. Solid understanding of one concept is required to successfully build to the next concept.

Talking & Writing Mathematically

You'll notice that the activities in this book include a great deal of mathematical dialogue and writing, not only for the teacher but also for students. If they're to become solid mathematicians, children need to be able to speak and write mathematically. In order to do this, they first need to hear correct mathematical language, read mathematical writing, and take part in the writing process in a mathematical context. Here are some pointers for encouraging mathematical communication in your classroom.

Encourage children to talk during math lessons. When children talk with one another, they're helping each other make sense of what they're learning and they're cementing their own understanding

When you want your class to talk mathematically, tell them that! Say, "Boys and girls, I need you to discuss this with your friends using math talk." That lets your students know that you're expecting to hear only math conversations, not talk about ballet class, soccer practice, or TV shows.

Demonstrate what "math talk" sounds like so everyone understands what's expected. Grab every opportunity to model for your students correct ways to write and speak mathematically. At first you'll need to look for opportunities: "Gee, is there a way I can add a little math talk here?" But with some practicing, we promise that talking mathematically will become second nature!

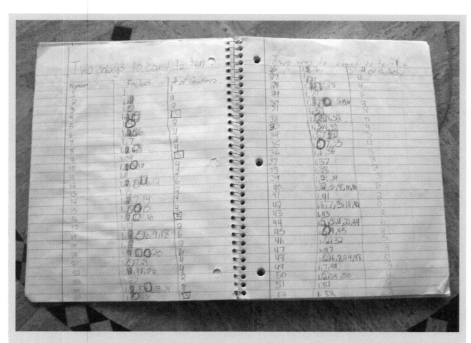

Math journals help students see how their own mathematical understanding grows through the year.

Encourage your students to keep math journals and to fill them with drawings, sketches, and diagrams (Whitin and Whitin, 2000). A simple spiral-bound notebook that begins its life totally blank can become a treasure of documented mathematical thinking and discoveries. As the year goes on, your students will become better at recording what they've learned—and that's what makes a math journal such a wonderful way to show growth over time. Imagine showing a parent how a child wrote about math in September and again in April! That is pure authentic assessment.

A WORD ABOUT VOCABULARY

There are many new vocabulary terms that students at each grade level are expected to learn. Correct use of mathematical vocabulary is critical for our students. They'll be expected to use correct terminology on CCSSM tests—both in multiple choice and in constructed responses.

Just as we use synonyms such as "protagonist/good guy" and "the climax/exciting part," in language arts class to help kids understand terms, we must use interchangeable language such as "decompose/pull apart" or "number model/number sentence" in math conversations too! In this book we deliberately provide many examples of interchangeable language. Expect correct vocabulary from your students and praise them when they use it. Not sure what all the correct terms are? Look them up in the math glossary in the back of your textbook or in the glossary on the Common Core website at www.corestandards.org.

Operations and Algebraic Thinking

Algebra has been known to make people shiver and cringe as they conjure up horrible memories of the mysterious Mr. X or Mr. Y. But it doesn't have to be that way. Algebra is so doable for kids if we let sound meaning and understanding sink in. Let's not be contributors of terror and trepidation to our students' algebra memory banks!

Children need to see that algebra has real-world application. They need context to construct meaning. The lessons provided here let your charges discover by doing. For example, third graders learn the commutative property by hopping around on colored cloths. Fourth graders discover numerical patterns by building an elaborate series of triangles using toothpicks. Fifth graders learn to see equivalency in complex equations by painting watercolor scenes.

The mathematics in this domain is rigorous, but if you take a careful look at the standards, you'll find that the progression of skills is very discernible, and the activities clearly transition from concrete to pictorial to abstract.

Third graders must create and explain patterns. They use manipulatives to solve two-step word problems, and they become fluent in multiplication and division up to 100. By the fourth grade, children are required to generate and identify features of patterns. They solve multi-step word problems, and they're responsible for graphing factors and charting multiples.

By the time our students leave the fifth grade, they must know how to analyze and compare two numerical patterns on a coordinate grid. They're also charged with solving complex, abstract problems involving algebraic symbols that include parentheses and brackets.

With this domain, you're truly opening the gate to more demanding algebra. Let's be sure this gate is welcoming and not mysterious or scary!

> Patterns are everywhere, and they can offer such powerful connections!

GRADE ③

Cluster 3.OA.A Represent and solve problems involving multiplication and division.

3.OA.A.1 Interpret products of whole numbers, e.g., interpret 5×7 as the total number of objects in 5 groups of 7 objects each. *For example, describe a context in which a total number of objects can be expressed as 5×7.*

Math Practices
7 Make Use of Structure
8 Express Regularity in Repeated Reasoning

Multiplication Masterpieces

 Paul Klee (Getting to Know the World's Greatest Artists) by Mike Venezia and

 Delicious: The Art and Life of Wayne Thiebaud by Susan Goldman Rubin

Examining a masterpiece or two of art each day not only sparks some culture and adds color to your math lesson, it also serves as a daily review of this standard.

To prepare, locate examples of works by Paul Klee, Wayne Thiebaud, or Wassily Kandinsky to share with your students. These artists all painted objects in arrays (Klee painted rectangles, Thiebaud food and other items, Kandinsky circles). Students will need tempera or watercolors, brushes and paper.

- Begin by displaying one piece of art at a time. When you show your students the first piece, ask them if they notice anything mathematical about the painting.

- Discuss how the painting shows images in rows and columns. Explain that the total number of images can be calculated by multiplying the number of rows by the number of images in each row. (For example, a work by Thiebaud may show food items in 5 rows of 7, illustrating $5 \times 7 = 35$.) Examine at least one to two works of art at the beginning of each math class for a few days.

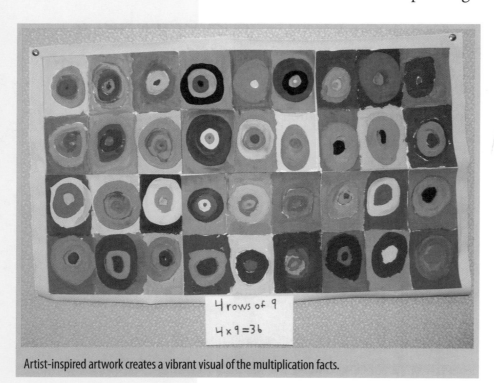

Artist-inspired artwork creates a vibrant visual of the multiplication facts.

- To manage the next step, you might like to set up a painting center in your classroom stocked with tempera or watercolor and let small groups rotate through the center.

- If your students are to paint in the style of Kandinsky or Klee, assign the dimensions for each row and column. (That way you can mount several student pieces together to show an even larger multiplication product.) For Thiebaud's food-inspired paintings, you'll want to assign a specific multiplication sentence to each child.

- Once displayed, the completed products provide your class with a variety of those critical multiplication sentences that need to be memorized.

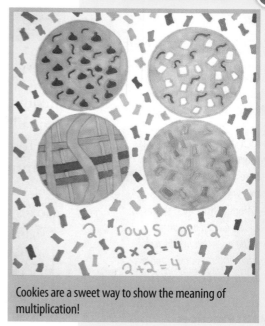

Cookies are a sweet way to show the meaning of multiplication!

Amazing Arrays

C P A
Individuals

3.OA.A.1

Math Practices
6 Attend to Precision
7 Make Use of Structure

In this activity kids create unique posters featuring four different models of one multiplication sentence.

Provide each student with a single-digit multiplication problem written on a small slip of colored construction paper (give each student a different problem). They'll also need an assortment of odds and ends (beans, stickers, paper cut-outs, etc.), a large sheet of white paper, markers, and glue.

Show students how to fold their papers to create 4 equal sections. Have them label each section as shown in the photo ("Groups," "Repeated Addition," "Fact Family," and "Array"). Tell kids to glue their single-digit multiplication problems to the center of their papers.

Pass out the markers, odds and ends, and glue. Instruct students to use the markers to fill in the "Repeated Addition" and "Fact Family" sections of their posters, and to use the glue and odds and ends to illustrate the "Groups" and "Array" sections. The finished posters illustrate each student's multiplication problem.

These visual models help students strengthen their understanding of multiplication.

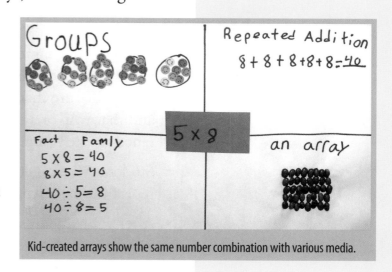

Kid-created arrays show the same number combination with various media.

3.OA.A.2 Interpret whole-number quotients of whole numbers, e.g., interpret $56 \div 8$ as the number of objects in each share when 56 objects are partitioned equally into 8 shares, or as a number of shares when 56 objects are partitioned into equal shares of 8 objects each. *For example, describe a context in which a number of shares or a number of groups can be expressed as $56 \div 8$.*

Math Practices
3 Construct Arguments & Critique Reasoning
4 Model with Mathematics

Divvy Up!

Division is the inverse of multiplication, correct? But take note that this standard does not address that mathematical property. Instead it requires students to focus on the fact that objects can be partitioned equally.

To prepare, gather small items like cubes, erasers, cotton balls, pom-poms, keys, beans, buttons, and blocks—the more variety the better. You can use lunch bags, berry baskets, plastic bowls, or shoeboxes to hold the materials. Student partners should have at least 50 small items in their container.

- Pass out the materials to each pair of students. One group may get a bag of lima beans, and another might get a bag of cubes. Later you'll rotate the different bags among the students.

- Say, "Okay mathematicians, reach into your bag and count out 18 items." Wait while the items are counted.

- Ask, "Can you take those 18 items and put them into 3 equal groups?" Give time for kids to arrange and double-check their answer.

- Let's say that your students who are working with the cubes are struggling. Suggest they take 3 of the 18 cubes and place those cubes in a line. Say, "Let's call this cube #1, this cube #2, and the last cube #3." Point to each cube as you say this.

- Next have the students place each of the remaining 15 cubes, 1 cube at a time, under the first 3 cubes. Ask, "How many equal groups did you create? Yes, 3. How many cubes are in each group? Yes, 6. So 18 divided by 3 is 6. Watch how I write this: $18 \div 3 = 6$."

- Continue this same procedure using different numbers. Let kids work with materials other than what they worked with in the first round. Be sure that you and your students say the division sentence and that you, or a student, write the corresponding division sentence on the board.

- Students won't "get it" with this one experience, so revisit and repeat the experience on other days.

- After your students have had ample opportunity to divide different numbers of objects into equal groups, pose this question, "When would someone want to divide a large group of objects into equal groups? Let's try to list times when it would be important to divide any-thing—people, shoes, candy—into equal groups."

- If necessary, provide an example. "The baker has 24 loaves of bread. He needs to divide the loaves evenly among the shelves of his bakery." Act as the scribe to record your students' ideas for partitioning objects into equal groups.

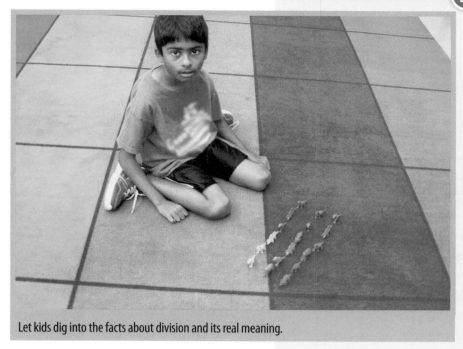
Let kids dig into the facts about division and its real meaning.

Write About It: Ask students to write a story in which the main character must divide a number of objects into equal groups.

Six-Dinner Sid

 Six-Dinner Sid by Inga Moore

Six-Dinner Sid is a charming story about a cat that eats 6 dinners each day! After you read this book to your students, they'll be ready to explore many mathematical possibilities.

For this lesson, children will need counters and grid paper. Students will find the grid paper useful for drawing diagrams of their concrete thinking. The paper also gives them a place to write their equations and label their illustrations as they attend to precision.

- After reading this story to your students, say, "Sid eats 6 dinners a day, right? So, if we know that he has eaten 24 dinners, can we figure out how many days have passed?"

- Ask students to take 24 counters. Say, "Sid eats 6 dinners each day, so arrange your 24 counters into groups of 6. How many groups do you have? Yes, 4. It would take Sid 4 days to eat 24 dinners." Explain to students that this problem can be written as "6 × ? = 24."

C ▶ P ▶ A
Whole Group, Pairs

3.OA.A.3 Use multiplication and division within 100 to solve word problems in situations involving equal groups, arrays, and measurement quantities, e.g., by using drawings and equations with a symbol for the unknown number to represent the problem.

Math Practices
1 Solve Problems & Persevere
4 Model with Mathematics
6 Attend to Precision

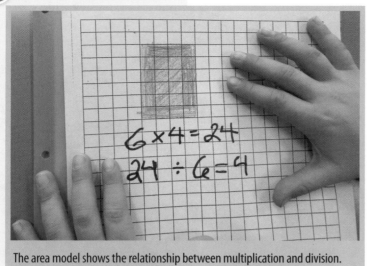

The area model shows the relationship between multiplication and division.

- Continue, "You can also solve this problem in a different way." Tell the students to arrange their 24 counters into an array with 6 rows. Explain that each column represents the 6 dinners Sid eats each day. Ask, "How many counters are in each row? Yes, it's still 4, but this time we used the equation "24 ÷ 6 = ?" to solve the problem."

- Now it's time to move students to the pictorial level. Pass out the grid paper. Invite kids to draw the 2 different models for solving this problem and write the matching equations. Remind students, "This problem can be solved as 6 dinners × an unknown number of days = 24 dinners, or you can think of it as 24 dinners ÷ by 6 dinners each day = the number of days."

- Next, you'll show your kiddos how to use a number line to solve this problem. You may need to project the number line, draw it on the board, or invite kids to sit very close to you so they can clearly see your demonstration.

- Demonstrate for students how to draw a number line on the grid paper. Explain, "Each little square will represent one number. Your line should extend 24 boxes to stand for the 24 dinners."

- Next, show children how they can jump 6 boxes forward at a time 4 times in order to arrive at 24. To model division as repeated subtraction, demonstrate how to start at 24 and subtract 6 four times to arrive at zero.

- Have partners create "Sid Problems" for classmates to solve. For example, "Sid has eaten 42 meals. How many days have passed?" Instruct each team to use equal groups or arrays and equations with a symbol for the unknown.

- After you check their work, let students exchange and solve each other's problems.

Extension: Pose this challenge, "How many days would it take for Sid to eat close to 100 meals without going over? You must use equal groups or arrays and equations with a symbol for the unknown." (96 meals would be 6 per day for 16 days. 6 × 16 = 96.)

Centipede's 100 Shoes

 Centipede's 100 Shoes by Tony Ross

In this story, after wearing his new shoes for several days, the little centipede decides that shoes are just too much trouble! He decides to give away his 100 shoes to friends with fewer legs. This plot provides your students with the perfect context to explore multiplication and division.

For this group exploration, each student will need 40 tiles, a 10 x 10-centimeter grid made using Dynamic Paper (see page 9), math journals or paper, and crayons.

- Read the story to your students once for enjoyment. The second time through, stop at the centipede's first gift of shoes to the 5 spiders. Lead a discussion about the number of legs on 1 spider.

C P A
Whole Group

3.OA.A.4 Determine the unknown whole number in a multiplication or division equation relating three whole numbers. *For example, determine the unknown number that makes the equation true in each of the equations $8 \times ? = 48$, $5 = \Box \div 3$, $6 \times 6 = ?$*

Math Practices
2 Reason Abstractly & Quantitatively
4 Model with Mathematics

The 10 x 10-cm grid provides children with an organized way to record the shoes as the centipede gives them away.

- Once your kiddos agree that spiders have 8 legs, pass out the tiles and say, "Let's use the tiles to show how many shoes went to the spiders." Encourage students to show 5 rows of 8 with the tiles.

- Distribute the 10 x 10-centimeter grid paper. Its 100 squares provide students with an organized way to keep track of the 100 shoes. Have students color the first 40 squares blue and create a key to show that blue represents the spiders' shoes. Ask kids to write the matching mathematical sentence "5 × 8 = 40."

- Next in the story, centipede gives shoes to four 6-legged beetles. Have students use tiles to model the number of shoes given to the beetles. They should also color 24 squares in the graph purple, add purple = beetles to the key, and write the matching number sentence.

- Woodlice are the next creatures to get shoes, but since most students are unfamiliar with them or how many legs they have, you'll want to save them for the last discovery.

- Continue with, "Friends, we've colored 64 of our 100 squares. There are still 36 shoes left. Grasshopper gets 6 shoes. How many shoes are left?" (30) Ask kids to use a yellow crayon to color 6 squares in the graph, add "yellow = grasshopper" to the key, and write the related number sentence.

- Now, the worms. Yes, 2 legless worms each get 1 shoe in this story! Have your kiddos color the 2 squares in the grid orange for the worms, add "orange = worms" to the key, and write the related number sentence.

- Say, "This leaves 28 shoes for the 2 woodlice, but how many legs does a woodlouse have?" Let students count and arrange 28 tiles in 2 rows. Continue, "Can you imagine a creature with 14 legs? Wow! Let's write this in the form of an equation."

- Say, "28 shoes left divided by 2 gives each woodlouse 14 legs." Write the equation "28 ÷ 2 = ?" on the board. Say, "We could also think of this as 2 woodlice times how many gives me 28?" Write the equation "2 × ? = 28" on the board. Have students finish by coloring the remaining 28 squares green and add "green = woodlice" to the key.

Write About It: Pose this problem, "Insects have 6 legs. How many shoes would 5 insects need? Show your thinking with an array and at least 1 equation." (Equations could be 6 × 5 = ? or ? ÷ 6 = 5.)

GRADE **3**

Cluster 3.OA.B Understand properties of multiplication and the relationship between multiplication and division.

Mix It Up!

This math standard doesn't require kids to know the definitions for the commutative, associative, and distributive properties, but it *does* expect kids to understand each term's meaning and know when to use it. These concepts are deep and shouldn't be hurried through.

The Commutative Property: You'll need 4 different-colored cheap plastic tablecloths from the dollar store. Cut the tablecloths in half and place the plastic tablecloth halves on the floor.

Let's say you use orange, red, purple, and yellow tablecloths. Call 4 kids to stand on the orange tablecloth and 4 more kids to stand on the purple one. Say, "Let's do some math talk. There are 2 sets of 4 children. We can write that as 4 + 4 or 2 × 4, correct? Okay, I'm going to rearrange these 8 children." Move 2 kids from the orange cloth to the red one, and move 2 kids from the purple cloth to the yellow one.

Ask, "How many children are on the tablecloths now? Right! There are still 8, but instead of 2 sets of 4, we now have 4 sets of 2, which we can write as 2 + 2 + 2 + 2, or 4 × 2." Write these expressions on the board.

Continue rearranging kids in this manner using different numbers. This activity seems to work like magic when it comes to helping kids understand the commutative property.

The Associative Property: Use a document camera, or ask students to gather on the carpet so they can watch as you demonstrate. You'll need 30 Unifix cubes.

Write the numerical expression "3 × 5 × 2" and ask, "What do you think about a problem with 2 multiplication signs?" Brace yourself. This type of problem will be brand new for most of your students.

3.OA.B.5 Apply properties of operations as strategies to multiply and divide. *Examples: If 6 × 4 = 24 is known, then 4 × 6 = 24 is also known. (Commutative property of multiplication.) 3 × 5 × 2 can be found by 3 × 5 = 15, then 15 × 2 = 30, or by 5 × 2 = 10, then 3 × 10 = 30. (Associative property of multiplication.) Knowing that 8 × 5 = 40 and 8 × 2 = 16, one can find 8 × 7 as 8 × (5 + 2) = (8 × 5) + (8 × 2) = 40 + 16 = 56. (Distributive property.)*

Math Practices
3 Construct Arguments & Critique Reasoning
7 Make Use of Structure

The number of children remains the same, but the arrangement and the multiplication sentence have changed!

There's so much power in a teacher-guided small group lesson when new concepts and procedures are introduced.

Introduce students to the distributive property. It'll be a part of their mathematics lessons for years and a strategy they can utilize for the rest of their lives.

QUICK TIP

You may certainly choose to use the terms "commutative," "associative," and "distributive" with your students; however, these terms won't show up on the standardized tests until middle school.

Eliminate the guesswork. Explain, "We know 3×5 means 3 sets of 5, so let's arrange 3 sets of cubes with 5 cubes in each set." Arrange the cubes and say, "That gives me 15 cubes. Let's look at the rest of the problem. It says *times 2*, so we need 2 sets of 15." Add another set of 15 cubes, arranged so that both sets are very easy to see. Ask, "How many cubes are there in all? Yes, there are 30 cubes. $3 \times 5 \times 2$ is 30."

Your next order of business is to prove the associative property. Say, "What do you think will happen if I change the order of the problem to $5 \times 2 \times 3$? Will we get the same answer?" Use the same 30 manipulatives to demonstrate this problem; it helps students "see" that the answer doesn't change.

Begin, "Let's make 5 sets of 2." Arrange the cubes and say, "Okay, there are 10 cubes. The equation says *times 3*, so we need 3 sets of 10." Arrange the remaining cubes and ask, "How many cubes in all?" (30) Point out that you were able to solve both problems with the same number of cubes because these problems are equal. The order doesn't matter.

Of course this one demonstration isn't nearly enough! Divide students into pairs, give them similar problems and manipulatives, and let them work through this discovery for themselves.

The Distributive Property: Snag 56 manipulatives and display the problem $7 \times 8 = ?$ Say, "Let's suppose that you don't know the answer to 7×8. Here's a strategy that can help you figure out the answer. To begin, you must first decompose, or break apart, one of the numbers. Let's decompose the 7 into 5 and 2. We can do that because $5 + 2 = 7$."

Continue, "Now instead of multiplying 7 times 8, we can multiply 5 times 8 and then we can multiply 2 times 8. We know that $5 \times 8 = 40$." Create an array showing 8 sets of 5 manipulatives. Say, "We also know that $2 \times 8 = 16$." Make a second array with 8 sets of 2 and place it away from the first set of 40. Continue, "Next we just add $40 + 16$. Let's see, one 10 and four 10s make 50. Next we add the 6 and we get 56. So now we know that $7 \times 8 = 56$!"

Dreams of Division

 Amanda Bean's Amazing Dream by Cindy Neuschwander

C P A
Whole Group, Pairs

3.OA.B.6 Understand division as an unknown-factor problem. *For example, find 32 ÷ 8 by finding the number that makes 32 when multiplied by 8.*

Math Practices
1 Solve Problems & Persevere
4 Model with Mathematics

Amanda Bean's Amazing Dream is a "must-have" book for every class-room. The opportunity to explore multiplication and division is beautifully displayed on every page. Windows, candy, baked goods, bushes, and tiles are arranged in rows, arrays, and groups. Your class could take weeks to explore the illustrations and enjoy the learning!

Each pair of students needs 50 small manipulatives (pasta wheel shapes and cotton balls are ideal), and 1.5-centimeter grid paper for recording arrays.

To begin, read the story to your class. Ask kids to help you retell the story. This lesson revolves around the situations described in the book.

- Begin, "Amanda's dream starts with 16 bicycle wheels. How many bikes would that be?" Tackle this first problem as a whole-group think-aloud.

- Say, "Since 1 bike has 2 wheels, we need to know how many groups of 2 we have. 16 wheels divided into groups of 2 will give us the number of bikes. That problem can be written like this: $16 ÷ 2 = ?$"

- Explain, "We could also think of this as an unknown-factor problem. 2 wheels per bike × *something* equal 16 wheels. That problem is written like this: $2 × ? = 16$."

- Pass out the pasta wheels. Ask partners to count out 16 pasta wheels and then give them time to work on the problem. They may first create 8 groups with 2 wheels in each group, or they might place the 16 wheels in 2 rows and find 8 in each row.

- Highlight both approaches and then pass out the grid paper. Let each student work with a partner; require partners to show their thinking concretely with the manipulatives, to record their work on the grid

Pasta wheels are the perfect tool for figuring Amanda's wheel problems of $2 × 8 = 16$ and $16 ÷ 2 = 8$.

A multiplication and related division problem sets the stage for this wooly scene.

paper, and to write the matching multiplication and division equations.

The book has numerous situations for exploration. Here are just a few to get you started:

⊙ If you counted 32 sheep legs, how many sheep would that be? (4 × ? = 32 or 32 ÷ 4 = ?)

⊙ A group of sheep has 40 balls of yarn. Each sheep in the group is holding 5 balls of yarn. How many sheep are in the group? (5 × ? = 40 or 40 ÷ 5 = ?)

⊙ Amanda sees 18 bicycle wheels whiz by. How many bicycles would that be? (2 × ? = 18 or 18 ÷ 2 = ?)

⊙ Amanda is looking at a large window on a building and sees 24 window panes in all. There are 6 rows of panes. How many columns of window panes are there? (6 × ? = 24 or 24 ÷ 6 = 4)

Variation: The pasta wheels can spark several different kid-created problems dealing with wheels. Just think of the possibilities: a tricycle has 3 wheels, a school bus has 6, a tour bus has 8, a city bus has 10, and a semi-trailer truck and Boeing 747 airplane each have 18 wheels!

GRADE ③

Cluster 3.OA.C Multiply and divide within 100.

Cheetah Math

 Cheetah Math: Learning About Division from Baby Cheetahs by Ann Whitehead Nagda

Division problems abound in this book about baby cheetahs living at the San Diego Zoo. Students will love the beautiful photographs, and you'll love the way the story's context gives kids a clear purpose and reason for learning about division.

Students will need at least 50 small manipulatives such as cubes or counters, plus base-10 blocks, colored pencils, and centimeter grid paper.

- After reading through the book once, revisit page 8. The trainer is preparing 18 ounces of formula to feed the 2 cubs. Say, "I'd like you to use the cubes to show me how much formula each cub will get." Kids should place 18 cubes into 2 equal groups.

- Next, ask your mathematicians to make an area model using the grid paper. Say, "Boys and girls, this model will be a pictorial representation of what you did with the cubes. Please draw a 2 x 9 rectangle on your grid paper." This is the purr-fect time for you to look over shoulders to be sure they're on track.

- Add, "Now please label your work with the related division and multiplication equations. 18 divided into 2 rows equals 9 in each row, and 2 rows of 9 equal 18. Those are the labels I should see on your grid." ($18 \div 2 = 9$ and $2 \times 9 = 18$)

- Say, "On page 10 we learn that at 6 weeks old, each cub drank a daily total of 15 ounces of milk over 5 feedings." Ask, "How many ounces of milk did each cub get at each feeding?" Repeat the same process. Have students model with cubes, create an area model on the grid paper, and record the related division and multiplication equations. ($15 \div 5 = 3$ and $5 \times 3 = 15$)

- Turn to page 18. Ask, "If the 2 cheetahs ate a total of 44 pounds of meat every day (and they each ate the same amount), how much meat would 1 cheetah eat in a day?" Many third graders can do this

C P A
Whole Group

3.OA.C.7 Fluently multiply and divide within 100, using strategies such as the relationship between multiplication and division (e.g., knowing that $8 \times 5 = 40$, one knows $40 \div 5 = 8$) or properties of operations. By the end of Grade 3, know from memory all products of two one-digit numbers.

Math Practices
1 Solve Problems & Persevere
4 Model with Mathematics

Remind children that labeling the diagrams is practicing using precision—an important math practice!

using mental math, but this problem allows you to introduce the base-10 blocks as a tool to figure division in a nonthreatening way.

- Tell students, "Please show me 44 using the base-10 blocks." (4 longs and 4 units.) Continue, "Now, make 2 rows with the blocks to represent the 2 cheetahs. Be certain that your 2 groups are equal." (Each row will have 2 longs and 2 units representing 22 pounds of meat for each cheetah.)

- Ask students to draw this base-10 block representation on the grid paper using colored pencils to make the individual blocks stand out, as shown in the photo. Children should also write the matching division and multiplication equations. (The equations are $44 \div 2 = 22$ and $2 \times 22 = 44$.)

This book is full of division stories about the 2 cheetahs. Don't feel it's necessary to race through this book all at once. You can spend weeks exploring and dividing. Enjoy!

GRADE **3**

Cluster 3.OA.D Solve problems involving the four operations, and identify and explain patterns in arithmetic.

Martha Sells Letters for $100

 Martha Blah Blah by Susan Meddaugh

Martha Blah Blah is a book that even adults find hilarious. Amazingly, a dog named Martha eats alphabet soup and the letters travel to her brain, giving her the ability to talk!

Students need paper and pencil. The problems here are meant to be solved using traditional algorithms and/or mental math, but if a student is struggling, offer tools such as manipulatives, number lines, and/or a hundred chart.

After the class enjoys the read-aloud, post the following chart and pose these alphabet-related problems featuring the story's characters. They'll get your class thinking about unknown quantities as well as using mental math and estimation. These tasks can be solved in a whole-class setting or in small groups.

Letters	Value of Each Letter	Total Value
A, E, I, O, U	$5	$25
M, R, T, H (These are the consonants in Martha's name.)	$6	$24
The other 17 consonants	$3	$51

Granny, who's responsible for soup letter production, left out the 5 letters: V, W, X, Y, and Z, plus all 5 of the vowels! Alf says those letters are worth $50. Martha says that's not reasonable. Who is correct and why?

➥ Martha is correct. $50 is too much money. Only 5 of the letters are worth $5 and the other 5 are each worth $3. (5 × $5 = $25 and 5 × $3 = $15; $25 + $15 = $40)

Lou bought 22 letters for $98. Name at least 1 possible combination of letters he could buy. (He can buy more than 1 of each letter.) Please write equations for each step as you solve this problem.

➥ Here's one possible solution: 10 vowels (10 × $5 = $50). The letters M,R,T,H (4 × $6 = $24). The consonants B, D, J, N, P, S, V, Z (8 × $3 = $24). $50 + $24 + $24 = $98.

Helen has $99 worth of $3 letters. How many letters does she have? Write an equation using a question mark symbol for the unknown number of letters.

➥ (33 letters) $3 × ? = $99 or $99 ÷ 3 = ?

Estimate how many of the consonants in Martha's name it would take to make $100. Explain your reasoning.

➥ One possible answer would be "fewer than 20" because the consonants in Martha's name are worth $6 each and $5 × 20 is $100.

Based on the letter values listed on the chart, write a problem of your own that uses more than one operation.

➥ One possible problem is shown in the photo.

"Martha bought 'Happy Birthday to You' for a friend's party. What did the letters cost?"

3.OA.D.9 Identify arithmetic patterns (including patterns in the addition table or multiplication table), and explain them using properties of operations. *For example, observe that 4 times a number is always even, and explain why 4 times a number can be decomposed into two equal addends.*

Math Practices
7 Make Use of Structure
8 Express Regularity in Repeated Reasoning

Practicing with Patterns

Where in the world would math be without patterns? This standard simply can't be accomplished in a few neat and tidy lessons. There are many helpful pattern strategies that students need to know and use. The introduction and explanation of patterns need not take long, but you should revisit pattern-based strategies often throughout the year. We can't stress that enough!

The hundred chart and addition grid are tools—yes, tools—that we believe *all* children should have in their math journals (they're available for free on the NCTM Illuminations website, see page 235). The patterns in our number system can be seen so easily with these visual tools. This lesson, however, focuses on the multiplication grid, the doubling strategy, and input/output charts.

When your students arrive at a mathematical generalization, post these nuggets in a special place in your classroom, leaving lots of extra room so you have space to add on throughout the year.

Multiplication Grid: Once students understand how to use a multiplication grid, they'll be ready to engage in some great mathematical conversations:

⊙ What do you notice about *all* numbers when they're multiplied by 1? (The numbers remain the same. That's called the identity property.)
⊙ What do you notice about *all* numbers when they're multiplied by 2? (They're all even.)

- What do you notice about *all* numbers that are multiplied by 4, 6, 8, or 10? (They're all even.)
- What happens when an odd number is multiplied by an odd number? (Products of odd × odd are always odd.)
- What about all numbers multiplied by 10? (They always have a 0 in the ones column.)
- What about the products when a number is multiplied by itself? Mathematicians call that a "square number," like 3 × 3, 4 × 4, 5 × 5, and so on. Where do you see those on the multiplication grid? (The products form a diagonal line or staircase, as many kids call it, going from left to right.)

Cool Kid-Approved/Teacher-Sanctioned Multiplying Strategies: Prove these strategies with simple manipulatives arranged in arrays that are doubled, doubled twice, or more.

# dogs	# legs
1	4
2	8
3	12
4	16
5	20
6	24
10	40
20	80

#spiders	# legs
1	8
2	16
3	24
4	32
5	40
6	48
7	56
8	64

#nickels	value
1	5¢
2	10¢
4	20¢
6	30¢
10	50¢
12	60¢
15	75¢
20	100¢
25	125¢

#100 legged leonard the legged organisms	legs
1	100
2	200
3	300
4	400
5	500
6	600
20	2000

The first two teacher-generated input-output charts inspired this child to create more charts!

- When multiplying a number by 2, simply double the number.
- When multiplying a number by 4, simply double the number, then double again.
- When multiplying a number by 8, double the number, double again, then double one last time.

Once students "get this," they can multiply numbers like 45 by 4 using mental math. 45 doubled is 90 and 90 doubled is 180. Sweet. (Works well for adults, too.)

Input-Output Charts: These charts are part of our day-to-day life. Look at the numbers on small snack bags for real-world data.

Begin, "Boys and girls, this bag of fruit snacks is 10 ounces. Let's create a T chart to show the number of ounces in several bags. I'll write 'Number of Bags' and 'Number of Ounces.' I'll underline that." (Talk through every detail involved in making this chart.)

Continue, "Let's start with 1 bag. I'll write '1' under 'Number of Bags,' the input column, and I'll write '10' under the 'Number of Ounces,' the output column. Now let's say I have 2 bags (write '2' in the input column). How many ounces will I have? Yes, 20." Kids will soon see the pattern of × 10. If kids are ready, ask, "How about a half bag of snacks? How many ounces would that be? Yes, 5! Good for you."

GRADE 4

Cluster 4.0A.A Use the four operations with whole numbers to solve problems.

C ▶ P ▶ A
Whole Group

4.0A.A.1 Interpret a multiplication equation as a comparison, e.g., interpret $35 = 5 \times 7$ as a statement that 35 is 5 times as many as 7 and 7 times as many as 5. Represent verbal statements of multiplicative comparisons as multiplication equations.

Math Practices
4 Model with Mathematics
5 Use Tools Strategically

Add Comparison to Your Skills

Comparison is a major addition to the fourth grade curriculum. In third grade, your students engaged in multiplication situations involving equal groups, arrays, and area. This year they must use comparisons to find unknown products, to determine unknown group sizes, and to figure unknown numbers of groups.

For this activity, students need small manipulatives such as cubes or counters, and their math journals.

After your kiddos have had lots of experience with the concrete manipulatives, move them on to the pictorial stage by having them represent/draw the problems on graph paper. These concept-building activities will prepare them well to handle these types of problems using mental math and abstract reasoning.

Of course, solving just one problem for each of the mathematical situations modeled below won't be nearly enough practice! Create additional problems based on these and use them frequently throughout the year.

Unknown Products: First you'll want to help your students establish an understanding of this new situation. Begin, "Take out 5 cubes and place them on your desk. What would you need to do to find out what twice that amount would be?" (Multiply by 2.) Have students model this by taking out another set of 5 cubes. Continue, "Show twice as many as the original 5 cubes, and write a matching equation in your journal." ($2 \times 5 = 10$)

Most students will easily understand this missing product problem. When they need to figure twice the amount of something they know to multiply by 2. This same reasoning applies to situations in which they need to know 3 or 4 or even 100 times the amount. Multiplying is the solution.

Unknown Group Size: You want students to grasp that in order to solve these types of problems they need to use multiplication in reverse and divide the whole to find the number in each set.

Number models, tiles as tools, and student-teacher conversations equal a productive math class.

Start, "I have 20 tiles, and that's 4 times as many as my original number. What was my original number of tiles?" Children can take 20 tiles and place them in 4 rows. Once they do this they'll see that 5 tiles are in each row, so 4 rows of 5 tiles make 20 tiles. Have students write an equation using a question mark in place of the missing factor. ($4 \times ? = 20$) The unknown group size can also be found by the inverse operation, division. ($20 \div 4 = 5$)

Unknown Number of Groups: You want your students to understand that they must take a total number and repeatedly divide it into sets to find the number of sets needed.

Explain, "Now I have 24 tiles. If I started with 8 tiles, how many times greater is my number now?" The equation for this problem is $8 \times ? = 24$. Using the tiles, children will see that it takes 3 rows of 8 to make 24, so that 24 is 3 times greater than 8.

DOMAIN
OPERATIONS AND
ALGEBRAIC THINKING

4

Whole Group, Pairs

4.OA.A.2 Multiply or divide to solve word problems involving multiplicative comparison, e.g., by using drawings and equations with a symbol for the unknown number to represent the problem, distinguishing multiplicative comparison from additive comparison.

Math Practices
1 Solve Problems & Persevere
2 Reason Abstractly & Quantitatively

Multiply—How & Why

 Amanda Bean's Amazing Dream by Cindy Neuschwander

This activity takes its structure directly from Table #2 of the original CCSSM document. The table shows how and why multiplication is used in different situations. We've found that sharing and discussing this table with our students helps them to recognize the many reasons for using multiplication too!

Third graders work on the types of problems found in the first 2 rows of the table. Fourth graders continue with these and add problems that deal with comparison. The problems used in the table here are inspired by the wonderful book *Amanda Bean's Amazing Dream.*

This is an abstract activity; however, if students are struggling, encourage them to use tiles or cubes to solve the problems.

- Display the table shown on the next page. Slowly read and discuss the problems as a whole class. You may prefer to approach the problems row-by-row, or to work them in columns. Either way is fine. This standard is all about the *variety* of problems our charges see.

- The columns in the table represent the 3 different positions where the unknown is found: Unknown Product, Unknown Group Size, and Unknown Number of Groups. The rows show the **representations of multiplication**: making Equal Groups, working with Arrays/Area, and making Comparisons.

- This is not a race. You want to be sure that your students really understand *what* they're trying to solve. Be sure to point out that the unknown is in different places in each number model.

- Once kids have had opportunities to discuss the correct number model/number sentence and the correct position of the unknown, followed by figuring out the answer, they should be ready to begin writing problems and their related number models on their own.

- Invite kids to write 1 multiplication sentence for each position of the

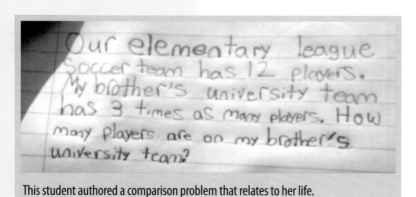
This student authored a comparison problem that relates to her life.

unknown. This isn't as easy as it sounds! Students may have to revisit the problems here for reference.

- Sometimes it helps kids to write problems if they have an "anchor" for their problems, such as hockey, dance class, soccer team, or the pizza parlor. When students self-select a topic that's interesting to them, they're usually more engaged in the work.

	Unknown Product $a \times b = ?$	Unknown Group Size $a \times ? = p$ and $p \div a = ?$	Unknown Number of Groups $? \times b = p$ and $p \div b = ?$
Equal Groups	There are 8 sheep and each sheep has 5 balls of yarn. How many balls of yarn are there in all?	If 40 balls of yarn are shared equally among 8 sheep, how many balls will each sheep get?	I need 40 balls of yarn. 5 balls of yarn are packed in 1 bag. How many bags will I need?
Arrays/Area	Amanda is at the bakery looking at rows of cookies. She sees 8 rows of 5 cookies. How many cookies are there?	If 40 cookies are arranged in 8 equal rows, how many cookies will be in each row?	If 40 cookies are arranged into equal rows with 5 cookies in each row, how many rows will there be?
Comparison	The candy store has 5 red lollipops. It has 8 times as many yellow lollipops. How many yellow lollipops does it have? Measurement: A piece of taffy is 5 cm long. How long will the taffy be when the machine stretches it to be 8 times as long?	The candy store has 40 yellow lollipops and that is 8 times as many as the number of red lollipops. How many red lollipops does it have? Measurement: A piece of taffy is stretched to be 40 cm long. That is 8 times as long as it was at first. How long was the taffy at first?	The candy store has 40 yellow lollipops and 5 red lollipops. How many times more yellow lollipops are there than red ones? Measurement: The taffy was 5 cm long at first. Now it is stretched to be 40 cm long. How many times as long is the taffy now as it was at first?

Whole Group, Small Groups, Individuals

4.OA.A.3 Solve multistep word problems posed with whole numbers and having whole-number answers using the four operations, including problems in which remainders must be interpreted. Represent these problems using equations with a letter standing for the unknown quantity. Assess the reasonableness of answers using mental computation and estimation strategies including rounding.

Math Practices
1 Solve Problems & Persevere
5 Use Tools Strategically

Problems with Seeds

 Anno's Magic Seeds by Mitsumasa Anno

Anno's Magic Seeds sets the stage for a fantastical tale of a seed that, when planted, yields 2 seeds each year. The author doesn't tell readers how many seeds are at the end of the story—the reader has to work that out on his own using all 4 operations. (Thank you, Mr. Anno!)

You might want to read this story during your literacy block and compare it to other folk tales or fairy tales (that way you'll also be touching upon CCSS.ELA-Literacy.RL.4.9). Ask your students to estimate the number of seeds on the last page. Record those estimates on a chart. It's always fun to go back to estimates after the answer is determined. Kids are thrilled if they were correct, or close to correct.

Then once math time rolls around, bring out the book again. This time you'll want to read very s-l-o-w-l-y to give your students time to decide what mathematical operation to use and to solve the problem on each page. You may want kids to work individually or in small groups.

Insist that students label their math equations by "Year 1," "Year 2," and so on. Explain that the equations and labels that they use must be clear and concise, so that if a teacher were beamed down from Planet Zarkon (explain to your captive audience that's the place where teachers read, write, and speak perfect English!), the teacher would know *exactly* what they were doing in their calculations.

Once the last page is read, give kids time to check their work. Compare answers and equations. Was there more than one way to arrive at the correct answer? (There almost always is!) Ask your students to explain their thinking and encourage fellow students to ask for clarification. (By the way, the answer is 7. But don't tell the kids.)

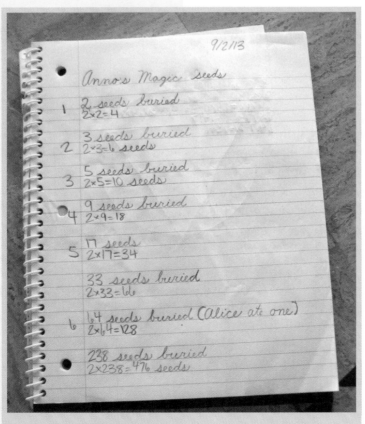

This journal page shows how multiple number sentences are needed to answer the one question, "How many seeds are there in the end?"

GRADE **4**

Cluster 4.OA.B Gain familiarity with factors and multiples.

More Than Two Ways

 Two Ways to Count to Ten, a Liberian Folk Tale by Ruby Dee

This story provides the perfect context for launching a discussion of factors, patterns, and prime and composite numbers. The bulk of this activity takes several days to complete, but students continue to add to their charts throughout the year.

You'll need chart paper and a marker. Each student will need a math journal or a sheet of paper and colored pencils.

- In this story, the animals of the jungle assemble for a challenge from the king—to throw a spear into the air and count to 10 before the spear falls to earth. Every animal fails except the ante-lope, who calls out, "2, 4, 6, 8, 10!" before the spear lands. The king is pleased and accepts him as the future king.

- After reading the book, you and your students will each create a "Ways to Count" chart. To begin, create an extended T chart with 3 columns. Label the columns "Number," "Factors," and "Number of Factors," as shown in the photo on page 38. Have the class follow along to create their own charts in their math journals.

- Next, list the numbers from 1 to 25 in the "Number" column (students will add more numbers later).

- Ask, "What are the factors of 1? Yes, just 1. Record that in the "Factors" column on your chart. Let's look at the number 2. What are the factors of 2? Yes, 1 and 2. Record those numbers on your chart." Continue with the numbers through 10.

- If your students feel capable (or if you know they are!), send them off to complete the "Factors" column up to 25. We try to list the factors through 25 on the first day and then add on another 20 numbers each day, until we reach 100. (Once your kids start to notice the patterns, it gets a lot easier!)

On the second day ask kids to gather their "Ways to Count" charts and colored pencils and get ready for a rich, non-rushed discussion.

Whole Group, Individuals

4.OA.B.4 Find all factor pairs for a whole number in the range 1–100. Recognize that a whole number is a multiple of each of its factors. Determine whether a given whole number in the range 1–100 is a multiple of a given one-digit number. Determine whether a given whole number in the range 1–100 is prime or composite.

Math Practices
7 Make Use of Structure
8 Express Regularity in Repeated Reasoning

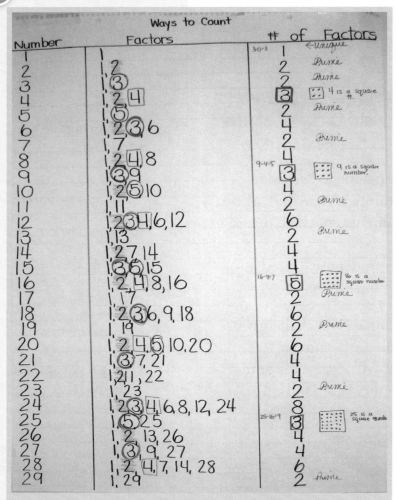

Whenever possible, remind your class about this mathematical fact with powerful charts like this.

- Ask, "Have any of you noticed a pattern in the factors? Did you notice that 1 is a factor for each number?" After ample discussion of their observations, say, "We're going to jazz up our charts so that all of the patterns you're noticing really pop!"

- Say, "Take a yellow pencil and highlight the number 2 every time it shows up as a factor." Be prepared for some "ah-has!" While they've known about even and odd numbers, it's probably never dawned on most of them that 2 is a factor in every even number.

- Say, "Use a red pencil and circle every 3. What do you notice?" (The numeral 3 appears in every third number.)

- Say, "Next make a green square around each 4. What do you notice? (The numeral 4 appears every other time a 2 is on the scene.)

- After the numbers 1 through 5 and 10 have been colored, your conversations should be filled with great generalizations (and your mathematicians will be working on math practices 7 and 8!).

- Continue, "Let's check out the factors for 12. If I multiply the first factor by the last factor (1 × 12), I get 12. If I multiply the second factor by the second to last factor (2 × 6), I get 12. If I multiply the middle factors of 3 and 4, I also get 12! Let's try this with the factors of 24."

You may want to fill in the third column on this second day, but don't feel that you have to rush this. This column simply lists the number of factors. Filling it in will go quickly. The reasoning will not come as quickly. Be patient.

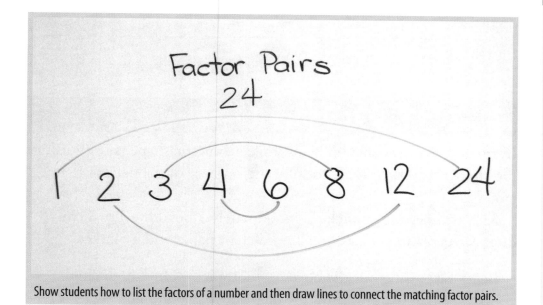

Show students how to list the factors of a number and then draw lines to connect the matching factor pairs.

- Direct students' attention to the "Number of Factors" column. Ask, "What's up with 1? It only has 1 factor." And "What about these numbers with only 2 factors." (They're prime.)

- After your conversation touches on many of the patterns the students notice, bring it to your students' attention that only the numbers 4, 9, 16, 25, 36, 49, 64, 81, and 100 have an odd number of factors. Ask, "What's special about those numbers?" (They're the square numbers.)

- Have kids focus on the difference (or distance) between each of the square numbers. Say, "From 4 to 9 it's 5, from 9 to 16 it's 7, and from 16 to 25 it's 9. What's going on there?" (The distance between one square number to the next increases by 2 and they're consecutive odd numbers.)

You'll find yourself and your students referring to this chart all year long, so hang this in a highly visible setting.

Write About It: Ask students to write about the patterns that they found on their charts.

C   P   A
Whole Group

4.OA.C.5 Generate a number or shape pattern that follows a given rule. Identify apparent features of the pattern that were not explicit in the rule itself. *For example, given the rule "Add 3" and the starting number 1, generate terms in the resulting sequence and observe that the terms appear to alternate between odd and even numbers. Explain informally why the numbers will continue to alternate in this way.*

Math Practices

7 Make Use of Structure

8 Express Regularity in Repeated Reasoning

GRADE ④

Cluster 4.OA.C Generate and analyze patterns.

Toothpick Triangles

In this activity, students build a simple geometric shape pattern using toothpicks, and then work to solve the mystery of the pattern that emerges.

Kids will need toothpicks (approximately 1 box for every 8 children), paper, and pencils. You'll need chart paper and a marker.

- Begin by passing out the boxes of toothpicks. Once each child takes a small handful from the box, you're good to go. Say, "Please create 1 toothpick triangle with 3 toothpicks." (So far, so easy.)

- Say, "We can all agree that 1 triangle used 3 toothpicks." Have students follow along as you demonstrate how to create a simple T chart. Tell them to label one side of their chart "Number of Triangles" and the other side "Number of Toothpicks." Have them write a "1" in the first column and a "3" in the second column.

- Continue, "Now, please add 2 more toothpicks so that you have 2 triangles that share a toothpick for one side. How many triangles do you see? How many toothpicks did you use? Add the answers to your chart." (2 and 5)

- Say, "Add another triangle. Now how many triangles do you have?" (3) "How many toothpicks did you use?" (7) Ask students to record their answers on their charts.

- Instruct kids to keep adding 1 more triangle in this fashion, up to 10 triangles. (At this point most kids will decide to work in teams because the toothpick lines

Touching and moving the toothpicks helps children see the growing pattern is not as simple as it first seemed!

become so long.) After each new triangle is added, pause and ask, "How many triangles do you have and how many toothpicks did you use?" Students should record their answers.

- Say, "Without creating another triangle, can you tell how many toothpicks you'd need for an eleventh triangle?" (23) "A twelfth?" (25) Those 2 problems should be fairly easy to solve. Kids will see that every triangle uses 2 more toothpicks and think "add 2."

- Once you feel that they've got it, ask, "So what if I didn't have the time or enough toothpicks to solve how many toothpicks I'd need for 34 triangles? Could there be a pattern that would help me figure this out?"

- Give kids time to study the input column (Number of Triangles) and the output column (Number of Toothpicks) on their charts. Resist the urge to give hints! It's good for the nine, ten, and eleven-year-old dendrites to work this out. What if they need to take this home and think about it? Fine! If kids become frustrated, encourage them to test their hypotheses. Ask, "Will doubling work? Will tripling work? Keep trying, I know you can figure this out."

- Once kiddos have had time to ponder independently, with partners, or in small groups, assemble the gang and listen hard to what they're expressing. Gently correct mathematical language, and encourage classmates to restate what's been said.

- Conclude, "Yes, if you take that number in the first column, the input, multiply it by 2, and then add 1, you get the number in the output column. Mathematicians express this as $2n + 1$. The n stands for the number of triangles." Suggest they try out this formula a few times to prove it. (For example, "$2 \times 6 + 1 = 13$, and I need 13 toothpicks for the sixth triangle.")

Write About It: Ask students to write about the patterns they discovered building the triangles and keeping track of the number of toothpicks.

Toothpick Squares

# squares	# toothpicks	
1	4	
2	7	
3	10	$3n + 1$
4	13	
5	16	
6	19	
7	22	
10	31	
15	46	
20	61	
50	151	
100	301	

In order to figure out the number of toothpicks needed for a number of square you will have to multiply the number of squares by 3 and add 1.

After kids have solved triangle toothpick problems, pose a similar problem about squares and number of toothpicks.

5.OA.A.1 Use parentheses, brackets, or braces in numerical expressions, and evaluate expressions with these symbols.

Math Practices
2 Reason Abstractly & Quantitatively
5 Use Tools Strategically

GRADE ⑤

Cluster 5.OA.A Write and interpret numerical expressions.

Counting Feet

 One Is a Snail, Ten Is a Crab by April Pulley Sayre

This standard introduces the mathematical punctuation of parentheses and brackets to our fifth graders. Because both of these symbols are new and kids need time to let these terms and symbols sink in, we've included 3 activities for this standard. This first activity operates on the pictorial level and uses parentheses only.

Students will need a sheet of 9 x 12-inch white drawing paper, a permanent marker, watercolors and a brush, their math journals, and pencils.

- Gather your students close, or use a document camera so they can see the illustrations in this whimsical book about counting feet. During the first reading, students will learn that a snail (a 1-footed creature) and a crab (a 10-footed creature) can be represented as 1 + 10.

- Next, using the animals from the story, model for your students how to write number sentences that use parentheses. For instance, say, "Watch how I'm going to write a number sentence to show how 60 feet could be represented by 5 crabs, 2 dogs, and 2 snails." Write: $(5 \times 10) + (2 \times 4) + (2 \times 1) = 60$.

- Explain, "There are 5 crabs; they each have 10 legs." Point to the (5×10). Continue, "There are 2 dogs and dogs have 4 legs." Point to (2×4). Say, "And then the 2 little snails each have 1 foot, so we write (2×1)."

- Model how to write several other number sentences that use parentheses involving characters from the story, and then invite your students to help you create additional number sentences using different animals and/or historical characters.

- Your kiddos may start by naming one type of animal, saying the number of feet that animal has, and then giving the total number of feet. (For instance 4 cows × 4 feet = 16 feet.) But you want your gang to move to using parentheses to hit this standard.

- You'll want to demonstrate a few more "feet problems" together before turning your students loose to create and solve 3 of their own "feet problems" involving at least 3 different animals or characters in their math journals. (You might require that the problems connect to your present science or social studies topics.)

- Once your author-mathematicians have written and solved their 3 problems, ask each child to circle the one problem he considers to be his most clever and marvelous.

- After checking their work, ask them to write another number sentence with a sum equal to the one in the problem they circled, but explain that this sentence must use different creatures and different number combinations. Have each student check in with you to make certain that their 2 equations are equal.

- Finally, have your kids copy and illustrate their 2 matching equations onto a sheet of drawing paper. We ask our students to draw in pencil first and then outline with a permanent marker before using the watercolor paints.

Displayed together, the students' work creates a wonderful bulletin board offering multiple examples of number sentences with parentheses! Powerful stuff!

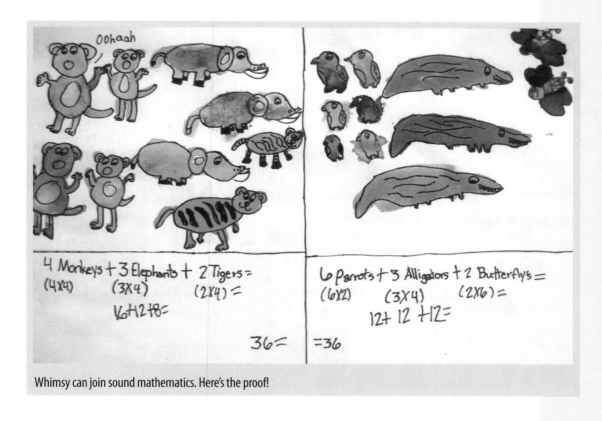

Whimsy can join sound mathematics. Here's the proof!

Individuals

5.OA.A.1

Math Practices

2 Reason Abstractly &
Quantitatively

3 Construct Arguments &
Critique Reasoning

Brackets

As problems become more complex, students will see brackets or braces. Some will be square-ended and others will be curly. Either way, brackets help students visually process an expression because they make each part of the problem more distinct.

Present your students with these problems: "9 × 2 + 3" and "9 + 2 × 3." Ask, "Are these equations equivalent? Why or why not?" This is a perfect time to practice constructing viable arguments!

Explain, "According to the Order of Operations we must always multiply first, so the answer to the first problem (9 × 2) + 3 would give us 18 + 3 = 21, and the answer to the second problem 9 + (2 × 3) would give us 9 + 6 = 15. Gee, without the parentheses, it's hard to keep things straight!"

Continue by saying, "Boys and girls, take a look at this problem: 9 × {7 × (6 +3)}. These new symbols are called brackets or braces and they help us see the problems more clearly. The brackets tell us to focus on the math equations inside the bracket first."

Say, "Inside the brackets we see parentheses, so first we do the math inside the parentheses, 6 + 3 = 9. Now look. What else is inside the braces? Yes, the 7. We must multiply the 7 by 9, which equals 63. Last, we multiply 63 by 9 and we get 567!"

Your students will need plenty of practice before this seems natural. Provide 2 or 3 problems like this each day as the mental math warm-up activity for at least a week.

Individuals

5.OA.A.1

Math Practices

2 Reason Abstractly &
Quantitatively

3 Construct Arguments &
Critique Reasoning

Decomposing Numbers

Once your Einsteins are comfortable composing answers by using parentheses and brackets, turn the tables on them! Ask them to decompose a number using parentheses and brackets! To begin, display and discuss this step-by-step decomposition for the number 24:

$$24$$
$$48 \div 2$$
$$(12 + 36) \div 2$$
$$[(3 \times 4) + (6 \times 6)] \div 2$$

Will G.

$$24$$
$$48 - 24 = 24$$
$$\{38 + 10\} - 24 = 24$$
$$\{[20 + 18] + [0 + 10]\} - 24$$
$$\{[(10 \times 2) + (2 \times 9)] + [(2 \times 0) + (2 \times 5)]\} - 24$$

This student understands that 24 is more than just 1 number after 23!

After walking your kiddos through each level of this decomposition, give your class a number and ask them to use parentheses, brackets, and braces to decompose that number. Encourage students to check one another's work. Compare the many different ways students can arrive at the same answer. This is great number sense practice!

Write About It: Tell each student to write a problem that uses parentheses, brackets, and braces and explain the order in which it must be solved.

Express Yourself

This standard requires students to write and understand simple numerical expressions that use parentheses. Our students aren't required to calculate an exact answer—what's important here is being able to interpret the numerical expression.

You can use the following activities with your entire class, with small groups, or as center activities. Let your students' understanding be your guide.

Match the Clues: This activity requires students to match a math expression with a clue. To prepare, create a simple chart like the one pictured on the next page for each small group of students. Have the kids cut out the expressions and clues. Students must then match the correct clue with its expression. Let each group share one match and justify its reasoning.

*Whole Group,
Small Groups*

5.OA.A.2 Write simple expressions that record calculations with numbers, and interpret numerical expressions without evaluating them. *For example, express the calculation "add 8 and 7, then multiply by 2" as $2 \times (8 + 7)$. Recognize that $3 \times (18932 + 921)$ is three times as large as $18932 + 921$, without having to calculate the indicated sum or product.*

Math Practices
2 Reason Abstractly & Quantitatively
7 Make Use of Structure

Mathematical Expression	Clue
$24 \div x = 8$	Triple 3 and then add 4.
3 (4 × 6)	Add 8 and 7 and then multiply by 2.
(3 × 3) + 4	Eight groups of 3
8 (3)	The quotient of 24 and x is 8.
2 × (8 + 7)	Three groups of 4 × 6

Is It Ten?: This activity can be used as a small-group activity, or as a formative assessment. Have students divide a sheet of paper into 3 columns. Label the columns "Less than 10," "Exactly 10," and "More than 10." Next post the expressions below and ask students to record each expression in the correct column.

Expressions: Answers:

(5 × 2) + 1   More than 10. This expression adds 1 to the 10.

1 × (5 × 2)   Equal to 10. This equation uses the identity property.

$\frac{1}{2}$ × (5 × 2)   Less than 10. The 10 is multiplied by a fraction less than 1.

$\frac{3}{2}$ × (5 × 2)   More than 10. The 10 is multiplied by a fraction greater than 1.

(5 − 4) × (5 × 2)   Equal to 10. The identity property is used here.

(5 × 2) × $\frac{1}{4}$ Less than 10. The 10 is multiplied by a fraction less than 1.

You'll notice that each expression has the equation 5 × 2 in parentheses. You want your students to think, "What's happening to the 10?" Again, you're looking for reasoning. An exact answer isn't required.

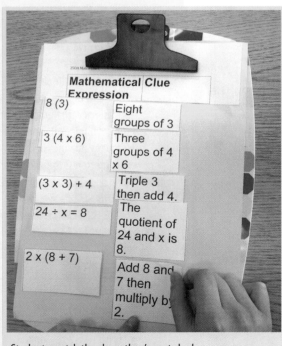

Students match the clues; they're not clueless.

GRADE 5

Cluster 5.OA.B Analyze patterns and relationships.

Patterns with Rules

Before children are ready to grapple with the abstractness of comparing two different rules on a coordinate grid, they must know how to use a coordinate grid! If that skill is new to your charges, don't worry—it's explained in detail here.

Students need a straightedge, a blank coordinate grid (this can be made using Dynamic Paper, see page 9), paper, and pencils.

The problems below are listed by increasing level of difficulty. Post just one problem a day.

Flossie and Betsy are rabbits. Flossie chews 2 carrots on day 1. Betsy chews 3 carrots on day 1. Each day Flossie chews 2 more carrots than the day before. Each day Betsy chews 3 more carrots than the day before. Create a separate T chart for the total number of carrots each rabbit chews in a week. Then plot the daily totals on a coordinate grid.

Luigi places 3 meatballs in every large bowl of his spaghetti. Mario places 5 meatballs in every large bowl of his spaghetti. Create a separate T chart for each chef, listing bowl by bowl the number of meatballs they would each need for 8 large bowls of spaghetti. Then plot the totals for the bowls on a coordinate grid.

Ava uses 4 beads in each necklace she strings. Riley uses 7 beads in each necklace she strings. Create a separate T chart for each jeweler, listing necklace by necklace the total number of beads needed for 10 necklaces. Then plot the totals for each necklace on a coordinate grid.

After your students complete their T charts for a problem, pass out the coordinate grids. Begin by pointing out each element of the coordinate grid for your students. The rabbit and carrot problem could go something like this:

- "This is a coordinate grid. Let's look closely at it. The numbers on the *x*-axis, or the bottom horizontal line, will represent the days on your T chart. Label your *x*-axis 'Days.' Now you need to fill in the

Individuals

5.OA.B.3 Generate two numerical patterns using two given rules. Identify apparent relationships between corresponding terms. Form ordered pairs consisting of corresponding terms from the two patterns, and graph the ordered pairs on a coordinate plane. *For example, given the rule "Add 3" and the starting number 0, and given the rule "Add 6" and the starting number 0, generate terms in the resulting sequences, and observe that the terms in one sequence are twice the corresponding terms in the other sequence. Explain informally why this is so.*

Math Practices
7 Make Use of Structure
8 Express Regularity in Repeated Reasoning

The points plotted and lines drawn on these coordinate grids open the gates to algebraic thinking and recording.

numbers. Place a '0' where the horizontal and vertical lines intersect. Next, write '1' for the first day, '2' for the second day, and so on."

- Continue, "The numbers on the *y*-axis, or the left vertical side, will represent the number of carrots the rabbits chewed. Write 'Number of Carrots' along the side of the *y*-axis." Explain, "Since the numbers must go up to 21, we'll label the *y*-axis in increments of 2."

- Of course, other grids may need to show numbers in increments of 5 or 10. Setting up the grids is a great opportunity to problem solve with your class. For that reason, we seldom use grids that have the numbers written on them. Deciding what numbers and what increments to use is part of the lesson!

- Demonstrate as you speak. "Let's plot the number of carrots Flossie chewed over 7 days. On day 1 she chewed 2 carrots. Put your right pointer finger on the 1 of the *x*-axis. Put your left pointer finger on the 2 of the *y*-axis. Push your fingers along the lines to meet. That exact point is called the coordinate. We write this as (1, 2)."

- Continue plotting all Flossie's numbers to Day 7. Once all points are plotted, say, "Now, take out a straightedge and pencil and connect the points."

- Repeat the above steps for Betsy's carrots on the same grid. Once the dots for Betsy's carrots are connected, discuss what the two lines show.

Remember, no matter what information the data represents, you always want to ask your students, "What generalizations can you reach about the two T charts and the coordinate grids?"

Number and Operations in Base Ten

Here's the plain and simple truth about this domain: It requires procedural knowledge and rote learning. At first glance this might seem almost contradictory to what we think of as the core of CCSSM—critical thinking, communicating, precision in language, and reasoning. But take a closer look at each standard.

Yes, students are expected to develop fluency (the ability to solve a problem quickly and accurately), but check out these verbs used in each NBT standard—*recognize, illustrate, compare,* and *explain.* Students are also required to understand the procedures and strategies used in solving problems. They should not just be mindlessly repeating steps.

When students say and use the steps required to solve a problem efficiently, they should be doing so because they've internalized the reason for each step. When kids understand the procedures they're using, and can articulate why they're using them, you can rest assured that they'll be able to apply their knowledge to the more complex problems that are on their mathematical horizons.

Once again, we suggest that you look closely not only at the standards for your grade level, but also at the ones before and after the grade you teach. Third graders who learn to add and subtract numbers to 1,000 and master basic multiplication facts to 100 will transform into fifth graders who multiply and divide multi-digit whole numbers and divide decimals. The leaps are huge!

We've deliberately selected many concrete and pictorial activities for this domain. Students would be well served if most of these activities were experienced more than once.

Whenever you can, seize the opportunities to connect your lessons to place value!

Whole Group

3.NBT.A.1 Use place value understanding to round whole numbers to the nearest 10 or 100.

Math Practices
4 Model with Mathematics
5 Use Tools Strategically

GRADE ③

Cluster 3.NBT.A Use place value understanding and properties of operations to perform multi-digit arithmetic.

Race Car Rounding

Teaching students how to round 2- and 3-digit numbers to the nearest tens place is fun and easy when you use a speed bump and race car!

For this hands-on activity, you'll need sentence strips, an empty oatmeal container, and a small toy car (such as a Matchbox car).

To prepare, label one sentence strip with the numbers 70 through 80. Place the 70 to the far left and the 80 to the far right. Underline all digits in the tens place (7 and 8).

- Begin by placing a "bump" (the oatmeal container) under the sentence strip beneath the number 75.

- Place the car on the number 77. While still holding onto the car, ask the class, "When I let go of this car, where will it roll—to the number 70 or to the number 80? Wherever it rolls is the nearest ten." Let go of the car and watch your students' reactions as the car rolls toward the 80.

- Ask, "Why did the car roll toward 80, not 70?" (80 is the closest 10. Gravity took over. The car would have to roll up and over the bump and then down again to reach 70.)

- Explain to students that when a number is rounded to the tens place, all of the digits to the right of the tens place (the ones place) are represented with a zero.

- Repeat this activity several times, changing the position of the car. Don't rush through this lesson. You're providing a powerful, kid-friendly visual, and kids benefit from repeated practice.

Now it's time to teach the rule about rounding up to the next tens place when the digit in the ones place is a 5.

- Place the bump under the number 75. Hold the car on the number 75. Ask the class where the car will roll.

Release the car and watch your students cheer as it rolls toward where they predicted. This is a great stick-in-their-mind visual!

- You hope they'll recognize that the car can't go because it's in the exact middle. Explain that when this happens, we must round up.

When your kiddos are ready, prepare other sentence strips with 3-digit numbers to practice rounding to the tens place (and later to the hundreds place). Remind your students that the car will roll to the number that will "round" it to the tens (or the hundreds).

 QUICK TIP

When students are rounding numbers, have them underline the digit to indicate the place value to which they're rounding. This way they can look at the digit to the right to determine if they need to round up or down.

3.NBT.A.2 Fluently add and subtract within 1000 using strategies and algorithms based on place value, properties of operations, and/or the relationship between addition and subtraction.

Math Practices

7 Make Use of Structure
8 Express Regularity in Repeated Reasoning

The Compensation Strategy

Consider this problem:

$$900 \\ -\,468$$

Admit it. Teaching a subtraction problem like this one can be a nightmare! It's not fun for you or the kids. Instead of torturing yourself and your charges with, "There's a zero in the ones place, so go to the tens column; there's a zero in the tens column, so go to the hundreds column…," teach this compensation strategy.

A simple number line labeled 0 to 15 (made from a 6-foot-long strip of adding-machine paper) and 2 children are all you need to demonstrate a concept that'll guarantee your students' success and provide them with a strategy they can use all of their lives.

• Roll out your number line and say, "I need one child to stand on the number 5 and another child to stand on the number 8." Have your helpers face the class. After your volunteers get into position, ask, "What's the difference between 8 and 5? What is 8 minus 5? Yes, it's 3."

• Ask each volunteer to move 2 spaces to her left (toward the number 15). Say, "Now we have one friend on the number 10 and one friend on the number 7. What's the difference between 10 and 7? Yes, 10 minus 7 is 3. The answer is 3 again. Neat!"

• Now ask your volunteers to move 6 spaces to their right. Announce, "We have friends on the numbers 4 and 1. What's the difference between 4 and 1? Yes, 4 minus 1 is 3. Hey, 3 again!"

• Do this several times so that your students see that as long as each child moves the same number of spaces in the same direction, the difference between the numbers (and the answer for the subtraction sentence for the two numbers) stays the same.

• You'll need to repeat this demonstration using other numbers. It's very important that your students see this with more than just an answer of 3! The point you're striving to get across is that the distance is the same whenever what is done to the subtrahend is done to the minuend.

- After you've driven that critical point home, display the following problem on the board:

$$900$$
$$-\ 468$$

- Those two zeros in the number 900 really make kids stumble, but take a deep breath because a problem like this is going to be fun for the first time. Really. Success is fun!

- Say, "I'm going to make this prob-lem a lot easier to solve. I'm going to take 1 away from each number so that the number sentence 900 take away 468 becomes 899 take away 467." Write that problem on the board next to the first problem.

- It's important to stress this next point. "Look, I took 1 away from both numbers. That means the distance, or the difference between the numbers remains the same because we did the same thing to each number. This is just like what our friends did as they moved around on the number line. Now this problem is much easier to solve!"

When students move the same number of spaces in the same direction on the number line, they're demonstrating that the difference between the numbers always remains the same.

There's no getting around the fact that students must know their basic number facts to arrive at fluency when subtracting multi-digit numbers. However, the compensation strategy can help students arrive at answers with more efficiency and accuracy. It's important that students understand this is not a trick; it's a strategy, a sound mathematical strategy.

3

Whole Group,
Small Groups

3.NBT.A.3 Multiply one-digit whole numbers by multiples of 10 in the range 10–90 (e.g., 9×80, 5×60) using strategies based on place value and properties of operations.

Math Practices

3 Construct Arguments & Critique Reasoning

4 Model with Mathematics

7 Make Use of Structure

Hundred Charts

This lesson asks students to skip count using the hundred chart. This experience helps students better understand multiplication problems that involve multiples of 10.

Each student needs a hundred chart (these can be made using Dynamic Paper, see page 9), and a colored transparent disc or other small marker. You can demonstrate this lesson using a document camera or interactive whiteboard.

- To begin, pass out a hundred chart and a transparent disc to each child. Say, "Boys and girls, place your disc on the number 10. Now move your disc as we count by tens. 10, 20, 30, 40, 50…100." Students move their discs from 10 to 20 and so on. This is simply skip counting!

- Continue, "You moved down 1 row for each 10, correct? How could you show me 5 groups of 10 or 5×10? (Students should move their discs 5 times, adding another group of 10 each time.) Ask, "Is $10 + 10 + 10 + 10 + 10$ the same as 5×10?" (yes)

- Ask, "How could you show 4×10 on your chart?" (Students use their discs to move from 10 to 20 to 30 and 40.) Ask the class if $10 + 10 + 10 + 10$ is the same as 4×10. You want to be sure your charges understand that multiplication is related to repeated addition.

- Say, "Since you've counted by 10s on your chart, it should be easy to count by 20s. Who can show me how to count by 20s using the chart?" (20, 40, 60, 80, 100, simply move two rows down at a time.) Ask, "How can I show 3×20 or 3 groups of 20?" ($20 + 20 + 20 = 60$ or 20, 40, 60)

- Challenge students to count by 30s or 50s and ask questions such as, "How can you show 3×30?" or "How can you show 2 groups of 50?"

The hundred chart helps students see how multiplication is related to repeated addition.

GRADE ④

Cluster 4.NBT.A Generalize place value understanding for multi-digit whole numbers.

Ten Times the Money

Money, which has been part of CCSSM since second grade, is the perfect tool to help students understand that in a multi-digit whole number, a digit in one place represents 10 times what it represents in the place to its right.

Each child needs play money in the denominations of $1, $10, $100, and $1,000 (at least 10 of each bill). They'll also need a place-value chart. Create the chart on your computer, or ask students to fold a sheet of paper into fourths and label it as shown here:

Thousands	Hundreds	Tens	Ones
$1,000 Bill	$100 Bill	$10 Bill	$1 Bill

- Begin by telling students to place a $1 bill in the ones place. Ask, "How much is 10 times that dollar?" They might say ten $1 bills or one $10 bill.

- Ask, "What place value do you see to the left of the dollar on your place-value chart? Yes, the tens or the $10 bill place."

- Once they've located the tens place on their charts, continue to question students, "What would you have to multiply the $1 bill by to equal $10?" ($1 × 10 = $10)

- Say, "Now put a $10 bill in the tens place. How many $10 bills do you need to make $100?" (10) Ask, "What do you have to multiply the $10 bill by to get $100?" (10)

- Continue in this manner. Ask students to place a single-digit number of bills in the ones, tens, or hundreds column of their place-value charts, and then have them place 10 times that number in the next column. Remind students that $1 is $\frac{1}{10}$ of a $10 bill, and a $10 bill is $\frac{1}{10}$ of a $100 bill.

C **P** **A**
Whole Group

4.NBT.A.1 Recognize that in a multi-digit whole number, a digit in one place represents ten times what it represents in the place to its right. *For example, recognize that 700 ÷ 70 = 10 by applying concepts of place value and division.*

Math Practices
4 Model with Mathematics
6 Attend to Precision
7 Make Use of Structure

Money helps students understand this standard!

- Get your kids talking! Ask questions such as, "Think about the reasoning you're using to compare one digit to the next greater digit." The essential understanding students need to grasp is that as they move to the left on the place-value chart each digit is multiplied by 10.

- When you feel confident in your students' understanding, ask, "How can you use what you know to tell me how many $100 bills it will take to make $1,000 dollars?" (10) "Can you write an equation for this?" ($100 × 10 = $1,000)

- If you're not able to find a play $1,000 bill, then it's fine to stop at the hundreds place. However, keep the thousands place on the place-value chart. It provides reinforcement that a thousand is 10 hundreds and it gives kids the opportunity to use repeated reasoning.

Write About It: Have students write the number 222 in their math journals. Ask them to explain how the 2 in the ones place is different from the 2 in the tens place and the 2 in the hundreds place.

Variation: Kids benefit from repeated practice using different tools. On another day, bring out the base-10 blocks and have students create a simple place-value chart with the headings "Hundreds," "Tens," and "Ones." Repeat the lesson above having students use the ones, longs, and a 100 flat in place of the money.

A digit in one place represents 10 times what it represents in the place to its right. That's the essential understanding of this standard.

Equal Shmequal

 Equal Shmequal by Virginia Kroll

The animals in this story want to play tug of war, but they can't figure out how to make the teams equal. This provides the perfect context for a math adventure. Get ready to talk about weight, number names, expanded form to the hundred-thousands place and more!

Every child will need a math journal or sheet of paper, colored pencils, and 1 place-value chart for each animal you choose to discuss. You can easily make the place-value chart (as shown in the photo on the next page) using your computer or you can instruct students to create their own using large sheets of paper.

- Read *Equal Shmequal* to your class one time purely for enjoyment. Then reread it calling students' attention to the animals' weights. The 7 animals in the story use a seesaw to compare how much they weigh.

- Tell your class, "Let's start with the weight of Bear, which is 136,412 grams. Your first job is to decompose this number and show it in expanded form on your place-value chart."

- Show students how to use their colored pencils to write the digits in order from left to right on their place-value chart (as shown in the photo). Using the different colors helps students delineate the different values of each place. Next ask students to fill in the number name.

- Continue, "Bobcat comes along next, weighing in at 21,991 grams. Can you say this number, write it in expanded form, and write the number name on the place-value chart?"

- Ask, "Does Bobcat weigh more or less than Bear?" (less) Continue, "How do you know?" (Bear's weight is in the hundred-thousands place, and Bobcat's weight is only in the ten-thousands place.)

- Depending on your students' needs, have kids fill in additional place-value charts using the weight information provided for the remaining animals. (Deer 67,299 g, Wolf 44,905 g, Rabbit 1,360 g, Turtle 907 g, and Mouse 50 g.) Those struggling with place value will find repeated practice using each animal very beneficial.

C P A
Whole Group

4.NBT.A.2 Read and write multi-digit whole numbers using base-ten numerals, number names, and expanded form. Compare two multi-digit numbers based on meanings of the digits in each place, using >, =, and < symbols to record the results of comparisons.

Math Practices
6 Attend to Precision
7 Make Use of Structure

- Once students have written the weights in both number names and expanded form, it's time to focus on greater than, less than, and equal. Turn to pages 18 and 19 of *Equal Shmequal*. Wolf and Turtle are on the seesaw. Ask, "Which one has the greater weight? How do you know?" (Wolf weighs more than 44,000 grams and Turtle is less than 1,000 grams.)

- Your conversations may remain just that, conversations. However, you may want to challenge students to take this information and turn their thinking into a math journal entry. If students are recording their thinking, encourage them to use the < and > symbols. (You'll find several more examples of weight comparisons on pages 20 to 21 and 24 to 25 of this picture book.)

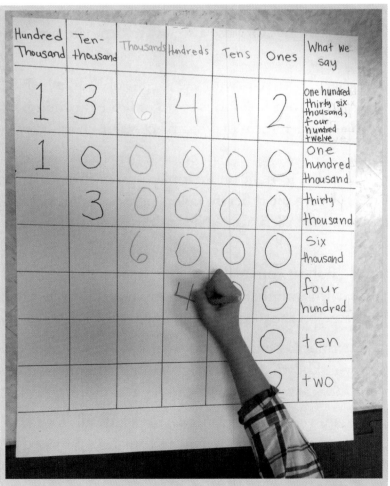

The colors the students use to fill in their place-value charts match the place-value strips described in Place-Value Rounding on page 59.

Place-Value Rounding

Place-value strips are excellent tools. They help children learn how to round numbers, understand place value, and compute large numbers. Math Practice Number 5 emphasizes the importance of knowing how to strategically select tools, so be sure to store your strips within easy reach. That way your students can grab 'em and go!

Place-value strips can be purchased online (crystalsprings.com) and at teacher stores, but they're also easy to make. Here's how:

Make Your Own Place-Value Strips: Cut ten 4 x $2\frac{1}{2}$-inch strips from paper and label each strip as indicated below using a thick, black marker. Numerals should be written so that when numbers are put together and the strips are stacked, only one number or one place-value position shows at a time as shown in the photo on the next page.

- ⊙ Ones: Cut white strips and label each strip with a number 0 to 9.
- ⊙ Tens: Cut red strips and label each strip 10, 20, 30, and so on to 90.
- ⊙ Hundreds: Cut orange strips and label each strip 100, 200, 300, and so on to 900.
- ⊙ Thousands: Cut yellow strips and label each strip 1,000, 2,000, 3,000, and so on to 9,000.
- ⊙ Ten Thousands: Cut green strips and label each strip 10,000, 20,000, 30,000, and so on to 90,000.
- ⊙ Hundred Thousands: Cut blue strips and label each strip 100,000, 200,000, 300,000, and so on to 900,000.
- ⊙ Millions: Cut purple strips and label each strip 1,000,000, 2,000,000, 3,000,000, and so on to 9,000,000.

- Begin, "All right, my friends, let's review rounding numbers. It's really helpful for estimating and for computing, so let's get solid on this concept."

- Display the number 2,478 using the place-value strips. Say, "Look at the colorful collection of numbers on these strips. This is 2,478. How many ones in that number?" (8) "How many tens?" (7) "How many hundreds?" (4) "How many thousands?" (2)

- Continue, "If I want to round this number to the nearest ten, I need to zoom in on the number in the ones place, correct? There's

4.NBT.A.3 Use place value understanding to round multi-digit whole numbers to any place.

Math Practices
1 Solve Problems & Persevere
5 Use Tools Strategically

Placing a sticky note, or sticky arrow, on the number you want kids to zoom in on helps them focus on the place value needing attention!

an 8. Since 8 is closer to 10, I round the 7 in the tens place to an 8, which means I'm rounding 78 to 80." Change the red color strip to 80 and the white color strip in the ones place to 0 to reassemble the number so it reads 2,480.

- Invite student volunteers to come up and practice rounding numbers to the tens place. Remind kids to zoom in on the ones place. If students seem confused by this review of rounding, they may benefit from the more concrete Race Car Rounding activity on page 50.

- Once you feel your class is ready, move on to rounding numbers to the hundreds place. Display the number 2,483 using place-value strips. Say, "If I want to round this number to the hundreds place, what place should I look to? Yes, the tens place."

- Say, "Let's see. There's a 4 in the hundreds place and an 8 in the tens place. That's 48 tens, or 480. Is 48 tens closer to 400 or 500? Yes, 500, so 48 tens, or 480 will round to 50 tens, or 500. The number 2,483 rounded to nearest hundred is 2,500." Don't settle for looking at only one numeral! Refer to the value of the entire number.

- You'll want to be sure your kiddos understand rounding to the nearest hundred before you venture to rounding to the thousands place.

Here are a few more ideas for using the place-value strips:

⊙ Pull out the place-value strips when your students are checking computation problems. Systematically ask someone to show the answer with the place-value strips and then round that number to the nearest ten, then hundred, then thousand, and so on.

⊙ Use the place-value strips whenever your class is working on estimation. Estimating is all about place value!

⊙ The place-value strips are the perfect tool when your students are faced with any independent rounding activity because the different-colored strips clearly show the different place values. Store them where your students can reach them!

Cluster 4.NBT.B Use place value understanding and properties of operations to perform multi-digit arithmetic.

Puzzle It Out

This standard is pretty cut and dried, rote, and procedural. Our fourth graders simply *must* be fluent with adding and subtracting multi-digit whole numbers. The tactics used in this activity add a bit of interest to those prickles of procedure.

Each student will need 11 sticky notes (or 2-inch scraps of paper) and a pencil. Instruct your students to write the numbers 0 to 9 on their papers, 1 number per paper. Have them write the plus sign on the remaining paper.

Challenge students to arrange their papers to create 2-digit addition problems, including the sums, as shown in the photo. Here's the tricky part: students may use each number only once! After your students have completed their work, record and check their problems with the class' help.

This activity can be used over and over again as a great mental math warm-up activity. Here are some variations:

- Ask kids if it's possible to arrange the sticky papers so they get an answer greater than 200. (No. Even with an 8 and 9 in the tens places, the sum will not be greater than 17 tens.)
- Challenge kids to find the highest possible sum. (94 + 82 = 176 and 92 + 84 = 176)
- Challenge them to find the lowest possible sum. (14 + 25 = 39 and or 15 + 24 = 39)

Whole Group, Individuals

4.NBT.B.4 Fluently add and subtract multi-digit whole numbers using the standard algorithm.

Math Practices
1 Solve Problems & Persevere
7 Make Use of Structure

This task works well for a wide range of ability levels because it doesn't require difficult math skills. It does, however, require kids to persevere and use critical thinking skills!

⊙ Switch the focus. Ask the class to create subtraction sentences.

⊙ Follow up by asking them to find the greatest possible difference (97 – 12 = 85) and the least possible difference (97 – 85 = 12).

⊙ Have your students create addition and subtraction sentences using 3-digit numbers. The same rules apply—students may use each number only once. Answers for subtraction problems may be 2- or 3-digit numbers. For example: 807 – 312 = 495.

Individuals

4.NBT.B.4

Math Practices

3 Construct Arguments & Critique Reasoning

7 Make Use of Structure

Think & Fill in the Blank

Solving the problems presented in this activity requires that kids think logically and systematically. They must use the rules of addition procedures to figure out the missing numerals. Post one problem as a quick warm-up activity for students to solve in their math journals before starting a math session.

The problems shown here are super-easy to create. Simply write and solve the problems when your students aren't in the room. Next, erase some of the numbers in both addends and the total. Just be sure you don't take away too much information, leaving it impossible to solve! When your kids enter the room, voilà, there are the problems ready to be solved.

These problems are very similar to the kinds of problems showing up on the CCSSM assessments.

Area Model Method—Multiplication

C P A
Whole Group

4.NBT.B.5 Multiply a whole number of up to four digits by a one-digit whole number, and multiply two two-digit numbers, using strategies based on place value and the properties of operations. Illustrate and explain the calculation by using equations, rectangular arrays, and/or area models.

Math Practices
1 Solve Problems & Persevere
7 Make Use of Structure

The area model method for multiplication requires kids to decompose or "pull apart" a number, understand place value, and employ basic multiplication facts. But once kids understand the concept, it's time to move on to teaching the traditional algorithm for efficiency.

- Begin with a very simple problem such as 5 × 35. Say, "The first factor has just 1 digit, so I'm going to draw a rectangle and write 5 on the left side."

- Continue, "Now, look at the next factor. The number 35 has 2 digits. Let's decompose 35 by place value. 35 is 3 tens or 30, and 5 ones. Since there are 2 factors, I'm going to divide the rectangle into 2 columns like this." Write 30 above one column and 5 above the other.

30   5

5 ☐ ☐

- Say, "Now, I'll multiply 5 times 30. Since 5 × 3 is 15, 5 times 3 tens is equal to 15 tens, or 150. So I'll write 150 in the first rectangle. Next, I multiply 5 times 5 and write 25 in the second rectangle. Last, I add up 150 and 25 and get 175!"

30   5

5 | 150 | 25 | 175

- Repeat this procedure a few more times, multiplying 1-digit numbers by 2-digit numbers. Emphasize place value each time.

When your students are ready, demonstrate how to use this model to multiply a 2-digit number times a 2-digit number.

- Say, "We're going to solve the problem 46 × 37 using the area model drawing method." Ask, "How many digits in the first factor? Yes, 2, so I'll draw a rectangle and then I'll draw a horizontal line dividing the rectangle in 2 parts."

- Continue, "Okay, now we must decompose the first factor. What does it become? Yes, 40 and 6. I'll write those numbers on the left side of the rectangle."

- Press on. "Now it's time to decompose the second factor, 37. It becomes 30 and 7. Since there are 2 digits, I need to divide the rectangle into 2 parts again. This time I'll draw a vertical line in the middle of the rectangle and write 30 above the first column and 7 above the second column."

$$46 \times 37$$

	30	7
40		
6		

- Say, "Now, it's time to multiply each number in this grid. Let's start with 40×30. Let's think about this. $4 \times 3 = 12$, so 4 tens $\times$ 3 tens = 12 ten tens. Yes, 12 ten tens. And 12 ten tens equals 12 hundred because 10 tens equals one hundred." Write 1,200 in the rectangle.

- You'll need to repeat that 10 tens equals one hundred, and 3 tens $\times$ 4 tens = 12 ten tens. You can take that to the bank!

- Next ask, "What's 40×7?" Break it down. "If $4 \times 7 = 28$ then 4 tens times 7 is 28 tens, or 280. I'll write 280 in the top right rectangle. Now, how about 6×30? $6 \times 3 = 18$, so 6 times 3 tens is 18 tens, or 180. I'll write that number in the bottom left of the rectangle. And, 6×7 is 42. I'll write that in the last box. Finally, we add up the 4 numbers and get 1,702.

	30	7
40	1,200	280
6	180	42

$$1,380 + 322 = \boxed{1,702}$$

- This is the prime time to show kids that no matter what order they add up the numbers, vertically or horizontally or even in a "z" design, the answer will always be 1,702. It doesn't even matter if students want to write in the 6×7 first. It's okay. That's the cool thing about multiplication and addition. Order doesn't matter!

Write About It: Ask students to explain why the area model method for multiplication works. (Students should say that breaking a number apart, or decomposing it, doesn't change its value.)

A Remainder of One

 A Remainder of One by Elinor J. Pinczes

In this story, the queen tries to divide 25 marching bugs into equal rows. This problem provides the perfect platform for the deep understanding required by this standard. As your students create rectangular arrays, they'll naturally link multiplication with division and address the question of remainders in a very nonthreatening manner.

Each student will need 25 tiles, plus base-10 blocks, 2 sheets of grid paper, and crayons.

- Begin by reading this story to your class once for enjoyment, and then go back to take a closer look at the math. The first arrangement the queen suggests is to have the 25 bugs march in 2 rows.

- Ask your problem solvers to show this arrangement using the tiles. Their tiles should show 2 rows of 12 with 1 tile remaining. This is a hands-on, concrete activity.

- Ask your students to move up to the pictorial level and record their tile arrangement on grid paper. Next, move your students to the abstract level. Ask them to write the matching equations, both multiplication and division, that describe their arrays. (Equations are $2 \times 12 + 1 = 25$ and $25 \div 2 = 12$ remainder 1.)

- Ask your Einsteins to continue their work following the same procedure using the queen's next three recommendations. Students will first divide the 25 bugs into 3 rows, and then they'll divide the 25 bugs into 4 rows. Each time, a remainder "remains." The final array pleases the queen because the bugs are in 5 rows with 5 bugs in each row, and there's no remainder to ruin the line.

C P A
Whole Group

4.NBT.B.6 Find whole-number quotients and remainders with up to four-digit dividends and one-digit divisors, using strategies based on place value, the properties of operations, and/or the relationship between multiplication and division. Illustrate and explain the calculation by using equations, rectangular arrays, and/or area models.

Math Practices
4 Model with Mathematics
6 Attend to Precision
7 Make Use of Structure

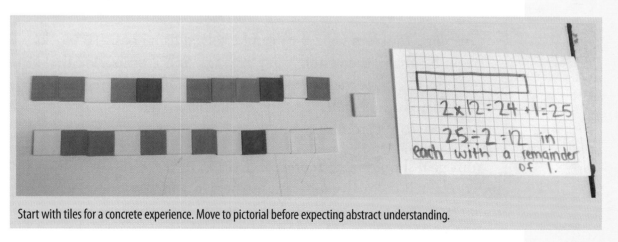

Start with tiles for a concrete experience. Move to pictorial before expecting abstract understanding.

✔️ **QUICK TIP**

Shodor.org offers math games that use interactive models and simulations. Remainders in Pascal's Triangle *at:* shodor.org/interactivate/ activities/ ColoringRemainder/ *is a fun way to practice with division and remainders.*

The area model provides a link to the work your students did as third graders involving area.

After the students have completed this easy entrance to division with remainders, they're ready to progress on to problems with 3- and 4-digit dividends.

- Ask students to create an area model for the equation 132 ÷ 6 using grid paper. Say, "How many columns would you need if you wanted to divide 132 by 6?" (22)

- The photo here models one child's thinking. She started with 6 rows. She knew that 6 × 10 columns would be 60 (as seen in orange). She added another 6 × 10 section (in green) for a total of 120 squares. She knew that 12 more squares were needed for the 132 in the dividend. The 6 × 2 columns (in yellow) gave her the required 132.

- This same problem (132 ÷ 6) can be solved using base-10 blocks. Begin by giving each child 1 flat, 3 longs, and 2 units. Ask the students to divide these pieces into 6 piles. Of course, regrouping (and therefore an understanding of the place-value system) is necessary to accomplish this task.

Your goal should be to represent division in many different ways in order to reach the many different learners in your classroom. Take as much time as needed for these kinds of division problems using arrays, area models, and base-10 blocks with similar problems.

GRADE ⑤

Cluster 5.NBT.A Understand the place value system.

Bring Out the Coins

Whole Group

This activity provides children with concrete and abstract experiences to ensure conceptual understanding of the place-value system. For students who would benefit from additional practice, repeat this lesson frequently in short doses.

To prepare, you'll need to gather a collection of pennies and dimes, and play $1 bills. For each student, you'll also need to make a simple place-value mat like the one pictured here.

- Begin by telling students to place a penny in the hundredths place. Ask, "How many pennies do you need to make a dime?" Reinforce that 10 pennies are equal to 1 dime or $\frac{1}{10}$ of a dollar and that the term "cents" comes from the Greek and Latin base word "centi" and means 100. There are 100 cents in a dollar. It's important to emphasize that the tenths place is the place to the left of the hundredths place on the place-value chart.

- Allow students the time to add 9 more pennies to the hundredths place, and then exchange that set of 10 pennies for 1 dime. Use the dime as a target point. Students need to understand that the penny/hundredths place is to the right of the dime/tenths place and that the penny is equal to $\frac{1}{10}$ of a dime.

- Next instruct students to exchange 10 dimes for $1. Students must understand that the dollar/ones place is to the left of the dime/tenths place, and a dollar is 10 times the value of the dime. You'll want to refer to this example of the concept several times until students' understanding of this standard is fluent.

- Continue to investigate this concept. Say, "Let's look at a different amount of money. How can you use the place-value chart to

5.NBT.A.1 Recognize that in a multi-digit number, a digit in one place represents 10 times as much as it represents in the place to its right and $\frac{1}{10}$ of what it represents in the place to its left.

Math Practices
4 Model with Mathematics
8 Express Regularity in Repeated Reasoning

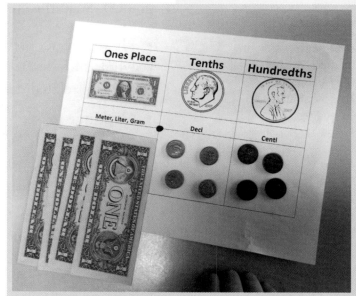

We like to add metric prefixes to our place-value charts. It's a good way to review and reinforce that metric prefixes are related to place value.

explain the difference in the values of the 4 in each digit for $4.44?"

- Starting with the 4 in the penny/centi/hundredths place, students should know that this digit represents 4 hundredths or 4 pennies. The 4 to the left is in the tenths place and represents 4 tenths or 4 dimes. The 4 in the ones place represents $4 or 4 times 10 dimes or 10 times the value of the ones place.

- Here's another way that students might explain the difference in values: 4 pennies is $\frac{1}{10}$ of 40 pennies or 4 dimes. The 4 in the ones place represents $4. $4 is 10 × 4 dimes and 4 dimes is $\frac{1}{10}$ of $4.

Write About It: Provide students with this problem, "Ben has $2.22. What is the value of each of the 2s in this amount?"

Whole Group

5.NBT.A.2 Explain patterns in the number of zeros of the product when multiplying a number by powers of 10, and explain patterns in the placement of the decimal point when a decimal is multiplied or divided by a power of 10. Use whole-number exponents to denote powers of 10.

Math Practices

3 Construct Arguments & Critique Reasoning

7 Make Use of Structure

8 Express Regularity in Repeated Reasoning

To the Point

This concrete activity demonstrates what happens when a decimal is multiplied by a whole number. In the past this tough concept was most often taught with the rote procedure, "Shift the decimal point to the left one place × 10, two places × 100, and three places × 1,000." But . . . why? This lesson answers that question!

You'll need 100 Unifix cubes or other snap cubes (if you don't have these cubes, borrow them from your teacher friends in the primary grades). Snap the cubes into 10 sets of 10.

- Hold up 1 set of 10 and say, "This set of 10 cubes represents the whole number 1. So 1 of the cubes is $\frac{1}{10}$ or 0.1." Write the fraction $\left(\frac{1}{10}\right)$ and related decimal (0.1) on a chart or board.

- Continue to ask questions such as, "What do 3 cubes represent?" $\left(\frac{3}{10}$ or 0.3$\right)$ and "What do 7 cubes represent?" $\left(\frac{7}{10}$ or 0.7$\right)$ Write these fractions and decimals on the board too. Once this understanding is established, display the problem "10 × 0.6" on the board.

- Ask, "What does 0.6, or six-tenths, of this set of 10 cubes look like?" Give students time to think. "Yes, it's 6 from this set of 10." Snap off 6 of the connected cubes and say, "Now think, the problem is 10 times 0.6, so what would that look like with these cubes?" Don't rush this! "Yes, it's going to be 10 sets of 6 cubes." Pull 6 connected cubes from each of the 10 sets.

- Continue, "Remember that 10 cubes snapped together equals 1 whole. So, what could I do with these 10 sets of 6 cubes to help me solve the problem?" Take a deep breath and pause. Patience will pay off. "Yes, let's take these 10 sets of 6 and snap them together to see how many sets of 10 we can create."

- Let your mathematicians digest this idea. At first it may seem foreign that 10 of something can represent 1 whole and that the sets of 6 cubes are going to be put back into sets of 10. Give students private think time. Tell them to write their answers in their math journals.

- Once all cubes are reassembled into sets of 10 your students will see that the 10 sets of 0.6 are now six sets of 10. Since each set of 10 represents 1 whole, the answer to 10×0.6 is 6!

- Will they "get it" after one demonstration? No way. So repeat with 10×0.5, 10×0.8, 10×0.9, and so on.

- Repeated practice with a variety of manipulatives is the key. Collect 100 pennies and 10 dimes and follow the same steps outlined above, using the dime to represent 1 whole (1.0) and each penny to represent 0.1.

Extension: To practice dividing whole numbers by decimals, repeat the activity in reverse. After all, division is the inverse of multiplication!

The Unifix cubes make it easy for kids to see that 10×0.6 is 6.

C P A
Whole Group

5.NBT.A.3 Read, write, and compare decimals to thousandths.

5.NBT.A.3a Read and write decimals to thousandths using base-ten numerals, number names, and expanded form, e.g., $347.392 = 3 \times 100 + 4 \times 10 + 7 \times 1 + 3 \times \left(\frac{1}{10}\right) + 9 \times \left(\frac{1}{100}\right) + 2 \times \left(\frac{1}{1000}\right)$.

Math Practices
4 Model with Mathematics
7 Make Use of Structure

A Milliliter Drop

When students place a milliliter-sized drop of water into their hands, it not only gets this lesson off to a good start, it provides an irreplaceable hands-on experience.

You'll need as many milliliter droppers as you can get your hands on (these are usually available for free at your local pharmacy). You'll also need several liter-sized containers (clean and empty water bottles work well) and a few jars of water.

Students need their math journals and a copy of the Place-Value Chart on page 220. This information-filled chart is worthy of being kept in your students' math journals all year long!

- To begin, distribute the droppers and jars of water. Say, "Taking turns with the droppers, place 1 drop of water on a friend's hand. Explain, "This dropper holds 1 milliliter of liquid, so the drop in your hand measures 1 milliliter."

- Have students look at a place-value chart. Say, "1 drop times 10 would give us 10 drops, and that would move us one place to the left on the place-value chart, to the centiliter place." Continue, "What do you think multiplying the 10 drops that represent the centiliter by 10 would give us? Yes, 100 drops, and that represents the deciliter."

- Add, "And if we multiplied the 100 drops times 10, how many drops of water would we have? Yes, 1,000 drops, and we'd land at the 1 liter place." Remind students that the base unit for each metric measure—meter, liter, and gram—is the ones place.

- You won't have the time (nor will your students have the patience) to completely fill a liter container with water 1 milliliter drop at a time; however, pass out the empty liter containers and give the kids a bit of time to add water to the empty container using the milliliter dropper. This will help build the benchmark of the magnitude of milliliter to liter.

Now that students have a sense of the relational size of the place-value positions, they can begin to practice using number names and writing numbers in expanded form. Review that a milliliter is $\frac{1}{1,000}$ of a liter and that it takes 1,000 milliliter drops to fill a liter container.

In this lesson, a milliliter drop of water promotes powerful understanding of relational size.

Remind students that the centiliter is $\frac{1}{100}$ of a liter and the deciliter is $\frac{1}{10}$ of the liter.

- On the board write, "Britt poured 2.345 liters of juice in the pitcher." Read the sentence aloud, pointing out that we say "and" for the decimal (two and three hundred forty five milliliters). Demonstrate how to write 2.345 liters in expanded form: $2 \times 1 + 3 \times \left(\frac{1}{10}\right) + 4 \times \left(\frac{1}{100}\right) + 5 \times \left(\frac{1}{1000}\right)$.

- Once your students are ready, offer them similar sentences that include decimals and instruct them to write the number names as well as the numbers in expanded form in their math journals. Quick problems like these make great mental-math warm-ups!

C P A

*Whole Group, Small
Groups, Individuals*

5.NBT.A.3b Compare two
decimals to thousandths based
on meanings of the digits in
each place, using >, =, and <
symbols to record the results of
comparisons.

Math Practices
4 Model with Mathematics
8 Express Regularity in
Repeated Reasoning

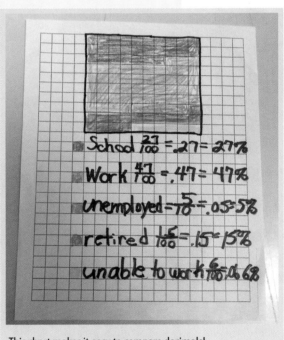

This chart makes it easy to compare decimals!

If America Were a Village

 If America Were a Village by David J. Smith

If *America Were a Village* presents facts and figures about our country
and its millions of citizens in a way that's easy to understand. The
author starts with the year 1900 and provides statistical information
that portrays what America would look like if it were a village with
just 100 people living in it. Over the course of the book, the author
updates the information in 20-year increments, bringing readers to
the present time.

Each student will need centimeter grid paper and crayons. You
might find a document camera helpful—it makes it easier for everyone to see the pictures as you read the story and demonstrate the
lesson.

- Introduce this book to your class, explaining that the
  author uses 100 people to represent the millions of
  people who live in America. Share information from
  a few of the pages with your class, and then open the
  book to page 17, which explores the occupations of
  the inhabitants of the village.

- Post the information that's provided on the page: Out
  of the 100 inhabitants, 27 people attend school, 47
  are workers, 5 are unemployed, 15 are retired, and 6
  are unable to work.

- Explain, "Each of you is going to create a 10 x 10
  grid to show the information presented here." Pass
  out the grid paper and crayons. Demonstrate how to
  outline a 10 x 10 grid at the top of the paper. Ask,
  "How many squares are in your grid?" Be sure students understand that their grids contain 100 squares.

- Say, "How many of the 100 people attend school?" (27) Continue,
  "Since you have 100 squares outlined on your paper, how many
  squares do you need to color to represent the 27 out of 100 people
  who go to school?" (27)

- Show students how to write the 27 out of 100 as both a fraction
  and a decimal. $\left(\frac{27}{100} \text{ and } .27\right)$ Instruct them to record this information on their papers below their grids.

- Continue in this manner, using the data for workers, unemployed, retired, and those unable to work. This is a great way to compare decimals because students can easily see the size of each decimal number represented on the grid.

- At the end of this activity you'll want the children to list the decimal numbers from greatest to least using the > symbol between each number. (0.47 > 0.27 > 0.15 > 0.06 > 0.05)

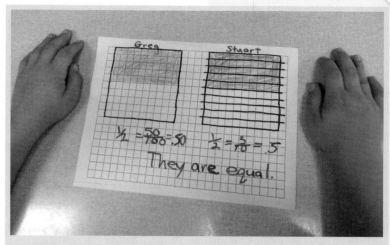

This side-by-side comparison of Greg's and Stuart's thinking helps demonstrate that the expressions 0.50 and 0.5 are equivalent.

Next, it's time to debunk a common misconception held by many fifth graders: 0.50 is *not* greater than 0.5. This activity is also a springboard for helping students think about decimals beyond the hundredths place.

Provide your students with a new sheet of grid paper and display the following problem. Your children might be ready to explore this problem on their own or in small groups, or you might decide to assign it as a whole-group task.

Stuart knows that about half of the people lived in towns and cities in 1920. Greg says that since the village has 100 people he can write that fact as $\frac{50}{100}$ or 0.50. Stuart says that he can also write that fact as 0.5. Greg does not agree. He says that 0.50 is greater than 0.5. Use 10 x 10 grids and words to prove who is right and why.

➡ Stuart is right. Both numbers are equal.

- If you're using the problem as a whole-group task, it's the perfect time to use two 10 x 10 grids side-by-side. Have students show Greg's answer by shading in 50 of the 100 small squares.

- Since Stuart's answer is 5 tenths $\left(0.5 = \frac{5}{10}\right)$, children must divide the second 10 x 10 grid into 10 equal pieces. Demonstrate how to draw horizontal lines to divide the grid into 10 rows of 10 as shown in the photo. When the children color 5 of the 10 equal sections, they can plainly see that the two expressions are equivalent.

- Push your students to the thousandths place by asking, "Which is the greatest? 0.5, 0.50, or 0.500?" (They're all equal.)

Whole Group

5.NBT.A.4 Use place value understanding to round decimals to any place.

Math Practices
4 Model with Mathematics
7 Make Use of Structure

Rounding Decimals

Rounding decimals is a new concept for fifth graders. The activities presented here are meant to lay the foundation for deep conceptual understanding.

For these activities you'll need meter sticks, base-10 blocks, index cards, and a box of flat toothpicks.

Round to the Nearest Decimeter: Gather kids around you and lay a meter stick flat on the floor. Place a base-10 long next to the meter stick. Say, "Look at how the long measures exactly 10 centimeters or a decimeter."

Ask a student to measure a unit cube along the meter stick. (It equals 1 centimeter.) Then, place 10 units along the stick to show 10 centimeters or 1 decimeter. Have a helper trade the 10 units for a long to represent the decimeter.

Now the rounding begins! Write the number 0.32 on an index card. Underline the 3 to help kids focus on the digit that you want them to round to. Ask, "How could we round 0.32 to the nearest tenth?"

Place 3 longs and 2 units along the meter stick to show 0.32. Ask, "Is 0.32 closer to 3 decimeters or 4 decimeters?" (3) Use this technique for similar questions until your students understand the meaning of rounding to the nearest decimeter.

Give your students time to practice rounding numbers to the nearest decimeter. Pass out the meter sticks and base-10 blocks. Remember to underline the digit in each number that you want your students to round to; it really helps them focus!

Round to the Nearest Centimeter: Fifth graders are charged with operating decimals to the thousandths place. Millimeters provide a concrete illustration to help students explore rounding even further.

Write 0.3<u>2</u>7 on an index card and underline the 2. Say, "How can we round 0.3<u>2</u>7 to the nearest hundredths place?" This is just like asking kids to round to

Toothpicks are perfect for marking the millimeter to facilitate rounding.

the nearest centimeter. Use a flat toothpick to mark 327 millimeters on the meter stick. A document camera makes it easier for kids to see each tiny slash mark on the meter stick, but if you don't have one, call the children close to you so that they can see this measurement up close. Ask, "Is the toothpick closer to .32 or .33?" Kids can see that it's closer to the 33 centimeters or 0.33 meters. You'll need to use multiple examples like this one to solidify understanding.

Once your students are comfortable with this new concept, instruct them to write "rounding to the hundredths place" problems for their peers on the index cards. Ask students to write their answers on the back of their cards. Check the cards for accuracy. Peers exchange and solve the problems using the toothpicks and meter sticks.

Write About It: Tell students, "Cathy has placed her toothpick on 0.555 on the meter stick. Please round this decimal to the nearest tenth, and to the nearest hundredth, and explain your thinking." (Rounding 0.5_5_5 to the nearest tenth is 0.6. Rounding 0.5_5_5 to the nearest hundredth is 0.56.)

GRADE 5

Cluster 5.NBT.B Perform operations with multi-digit whole numbers and with decimals to hundredths.

No Problem!

This standard has fluency and procedure written all over it. There's no getting around the fact that students must be able to fluently multiply multi-digit numbers. At the same time CCSSM want kids to think critically.

When your students first encounter the problems in this activity, they may wonder, "Is there something wrong with these problems?" That's asked because certain digits are missing, but that's precisely what makes these problems so powerful. Kids must use their multiplication skills and their number sense to solve them!

Completing problems like the ones provided here proves understanding of the standard. These problems are also very similar to ones showing up on the CCSSM assessments.

Whole Group

5.NBT.B.5 Fluently multiply multi-digit whole numbers using the standard algorithm.

Math Practices
2 Reason Abstractly & Quantitatively
6 Attend to Precision

Your students will need paper and pencil, and you'll need a location to display this problem:

$$
\begin{array}{r}
4\;5\;3 \\
\times\;\square\;9\;\square \\
\hline
3\;1\;7\;1 \\
\square\;0\;7\;\square\;\square \\
+\;\square\;7\;1\;8\;\square\;\square \\
\hline
3\;1\;5\;7\;4\;1
\end{array}
$$

Begin by asking, "What belongs in the squares?" If there's deadlock, suggest, "Take a look at the last square in the second factor. Do you see that the product has a 1 in the ones place? What could we multiply by 3 (point to the last digit on the first row) that would result in a number that has 1 in the ones place?"

Give students plenty of think time before you continue. "Yes, 7 would work. 7 × 3 is 21. Is it 7 for sure? Are there any other numbers that can be multiplied by 3 to come up with a product with 1 in the ones place?" (no) "Since we know that the 7 belongs in this place, can we solve more of the problem?" Students should be able to solve the rest of the problem, moving slowly and double-checking their work. Post more problems like this one and use them as mental math warm-ups before starting your math lesson. (The answer to this problem is 453 × 697 = 315,741.)

Mathematical conversation helps spark great thoughts.

For continued practice, ask your students to come up with the problems! Creating these kinds of problems is excellent practice. Students must be very judicious when deciding which numbers to remove and replace with a square. They must leave in enough information so the problem-solver can decide which numbers fit in.

Variation: Got struggling kids? Use place-value strips (see page 59) to decompose/pull apart the numbers. 675 × 45 is also (675 × 40) + (675 × 5). Using the place-value strips may help your students understand the necessity for the zeros.

Extension: If your charges have Internet access, have them research data-related facts such as, "How many crayons does the Crayola company produce in a day?" or "How many Lego pieces are manufactured in an hour?" Next have kids use multiplication to determine how many pieces of each product are created in a week/month/year and so on.

Go Figure, Father Time

Math problems imbued with meaning are much more engaging. Instead of asking your kids to multiply and divide numbers just for the sake of multiplying and dividing, why not give them something really interesting to think about?

You'll need to create and post a reference sheet with this information: "There are 60 seconds in a minute, 60 minutes in an hour, 24 hours in a day, 7 days in a week, and 365 days in a year."

This standard assumes that children can figure the answers to the questions posed here using paper and pencil and traditional algorithms. However, if some of your students aren't solid with the computation, you may permit calculators for the more complex computations. This way you're allowing the problem-solving element to be nurtured.

You'll notice that the questions in this activity are open-ended. They don't ask for the answers to be given using a specific unit of time. By asking only, "How long ago…" there's potential for diverse answers and rich classroom discussion. If you asked "How many minutes ago…" there would only be one correct answer.

- Begin with a review of the reference sheet and then ask, "How long ago was 180 seconds?" (3 minutes will probably be your most common answer.) Follow up with, "Why did you divide 180 by 60?" (There are 60 seconds in 1 minute.)

- Next ask, "How long ago was 420 minutes?" ($420 \div 60 = 7$ hours, which is easy for students to relate to. It's also 7/24 of a day, a bit less than a third.)

- Press on. "How about 550 minutes?" Since $550 \div 60 = 9$ with 10 left over, the answer most kids will get it is 9 hours and 10 minutes. Life isn't always neat and tidy. Dealing with remainders is a

C P A

***Whole Group, Small
Groups, Individuals***

5.NBT.B.6 Find whole-number quotients of whole numbers with up to four-digit dividends and two-digit divisors, using strategies based on place value, the properties of operations, and/or the relationship between multiplication and division. Illustrate and explain the calculation by using equations, rectangular arrays, and/or area models.

Math Practices
6 Attend to Precision
7 Make Use of Structure

✓ **QUICK TIP**

We believe distinguished educator Dan Meyer, a man known to create problems that generate thinking, would embrace this activity. If you're interested, look him up on YouTube, Ted TV, or on his blog at blog.mrmeyer.com.

fact of life, so let's be sure to include problems like this in our students' math experience.

- Once they've had practice with the concept, give them the 4 powerful problems below. These problems may seem deceivingly simple to your students at first. Your kids may be thinking, "Seriously, how tough can a 5-word problem with only 1 number be?" But each one will make your students stop and think: "What operation should I choose?" and "What step should I take first?"

⊙ How long ago was 2,400 seconds?
⊙ How long ago was 2,400 minutes?
⊙ How long ago was 2,400 hours?
⊙ How long ago was 2,400 days?

- Encourage children to show their work. No matter how kids choose to solve a problem, their solutions should be easy to follow. Some students will solve with traditional algorithms, some may draw arrays, and others might use the area models.

- Once problems are solved, call a "Math Meeting" so that students can take turns showing classmates their strategies and explain their reasoning. Insist on precise language in their descriptions and, if they used drawings, require labels. This is a perfect time to celebrate the different ways of arriving at the same correct answer.

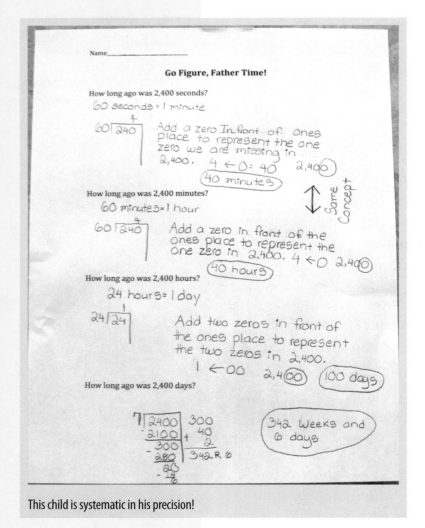

This child is systematic in his precision!

May I Take Your Order?

Individuals

Put out the word that you collect carry-out menus! Stash the menus you receive from your students and their parents near your math tools. There are endless possibilities for real-world mathematics within the pages of each one.

The prices on the menus, almost always shown in decimal form, can become fodder for some terrific problems. A menu-generated math warm-up one day a week can offer real-world practice with all operations. Some children may like/need to use paper money to solve these problems.

You're the boss, so tailor your questions to match the individual needs of the differing abilities in your classroom. Give menus with heftier price tags to children who need more challenges and menus with less complicated prices that are easier to read to those who are struggling.

There's no rule that everyone has to get the same problem, either. But, since your students are fifth graders, be certain you're providing them all with multiple-step problems.

Here are just a few ideas in order of increasing complexity to get you started:

- Figure the total cost for your family to eat a full meal at one of the restaurants.
- Find the difference between the most expensive and least expensive meal at three different restaurants.
- You have $134.50 to buy a complete meal for 10 people. What can you purchase? How much change will be left?
- Order 5 desserts, each a different price, for you and 4 friends. What's the total? If you split the bill evenly, what does each person owe?
- Find a delicious 3-course meal. Total all 3 courses, and then figure the cost for that meal for all the girls in the class, then all

5.NBT.B.7 Add, subtract, multiply, and divide decimals to hundredths, using concrete models or drawings and strategies based on place value, properties of operations, and/or the relationship between addition and subtraction; relate the strategy to a written method and explain the reasoning used.

Math Practices
3 Construct Arguments & Critique Reasoning
6 Attend to Precision

Three customers compute the tabs for their meals.

the boys in the class, and last for each teacher in our school. What would the total be if you added all 3 groups? What's the difference between the girls' total and boys' total?

⊙ You have 17 nickels, 37 quarters, 24 $1 bills, fifteen $5 bills, thirteen $10 bills, and 4 $20 bills. How can you spend all that money?

⊙ Select a different full-course meal for yourself and 4 friends. Once you have the total for the 5 of you, divide the bill evenly.

Students love to come up with problems for the menus. Solicit your class for ideas to add to your problem bank.

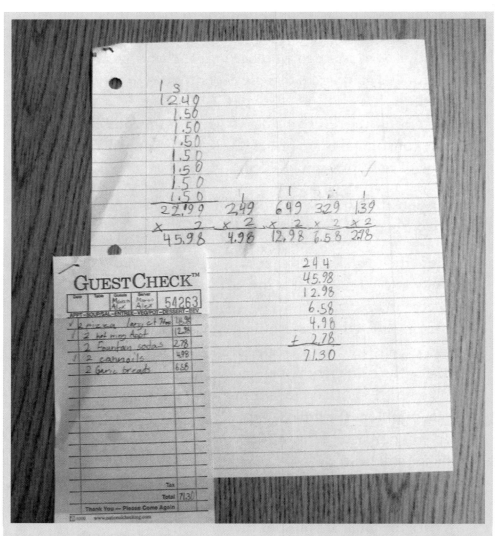

Swap lined notebook paper for simple green restaurant receipts and the task becomes more real-world play than work!

Number and Operations— Fractions

When you think of three quarters, what comes to mind? Music, minutes, how much gasoline is left in your car's tank? It's critical for our students to develop a solid and flexible understanding of fractions. Not only are fractions an important part of everyday life, but research shows that a child's ability to handle fractions in the intermediate grades has a direct correlation with his success in Algebra 1 and Algebra 2 (Gomez, 2009).

CCSSM expects third graders to understand and use fractions with the denominators 2, 3, 4, 6, and 8. Fourth graders add 5, 10, 12, and 100 to that list. Fifth graders are expected to work with unlimited denominators. Yes, *unlimited!*

Students need ample opportunity to work with and talk about fractions. Look for opportunities throughout the day to talk to your students in fractional terms. For example, when taking lunch count, you could say, "In this group, $\frac{5}{8}$ of the children are buying their lunch in the cafeteria and $\frac{3}{8}$ of the children are bringing their lunch from home." The more our students have the opportunity to work with fractions the better!

Students also need a wide range of models when learning fractions. We must move beyond pizzas and candy bars. Students need to see fractions as part of a circle, square, or rectangle; but they also help children to see fractions as part of a set, on a number line, and in everyday situations.

The activities in this domain help students understand that fractions transfer beyond the classroom into the real world. They also help prepare students for the more complex algebra that they'll encounter later on. So let's put our whole hearts into teaching these important parts!

Students need ample opportunity to work with and talk about fractions.

GRADE ③

Cluster 3.NF.A Develop understanding of fractions as numbers.

3.NF.A.1 Understand a fraction $\frac{1}{b}$ as the quantity formed by 1 part when a whole is partitioned into b equal parts; understand a fraction $\frac{a}{b}$ as the quantity formed by a parts of size $\frac{1}{b}$.

Math Practices
2 Reason Abstractly & Quantitatively
3 Construct Arguments & Critique Reasoning

My Friends & Me

Call 4 children to stand up in the front of the room. Say, "This is 1 group of children. There are 4 children in this group. One of the children in this group has a ponytail. She is 1 of 4 in the group, so I can say $\frac{1}{4}$ of the children in this group has a ponytail."

Say, "$\frac{1}{4}$ looks like this: 1 over 4. (Write '$\frac{1}{4}$' on the board.) We say it like this: 'one fourth.' The bottom number is called the 'denominator' and the top number is called the 'numerator.' The denominator tells how many parts there are in the whole group. The numerator tells how many parts we are talking about."

Encourage students to talk about the different attributes of the children who are in the group using fractions. The possibilities are endless—collars, color of pants, shoes with laces or Velcro, pictures on sweatshirts or shirts, buttons or not, eye color, hair color, gender, or favorite sports teams.

Students love this simple activity and it's worth repeating throughout the year. Be sure that you stick to the CCSSM denominators that third graders are expected to fully understand (2, 3, 4, 6, and 8).

When, or if, you feel that your students are ready, you may want to move on to other denominators.

Extension: Ask each child to color and cut out a self-portrait. Place children, and their paper likenesses, in groups of 2, 3, 4, 6, or 8 (these are the very numbers that third graders are to use as denominators). Ask each group to come up with 5 fraction sentences about the portraits in their group (see photo on page 83).

This activity is a great way to get students thinking and talking in fractional terms.

Kids write about themselves using fractions!

Paper-Folded Fractions

C P A
Small Groups

3.NF.A.1

Math Practices
4 Model with Mathematics
6 Attend to Precision

This is such an easy-to-do lesson, yet the results are so powerful. By folding paper strips, children create concrete models to show equivalency among fractions with different denominators (specifically, halves, thirds, fourths, sixths, and eighths).

Each student needs 5 paper strips that each measure 2 x 10 inches. You may choose to assign each fraction its own separate color, but that's not necessary.

- Pass out 1 strip of paper to each student. Say, "We're going to create and label fractions today. We're going to be as precise as we can in our folds and we're going to label the fractions. Folding can sometimes be tricky, but we'll persevere and do our best. I'll go step-by-step. You'll need to watch each step and then follow the fold I make, just like follow the leader. We're going to take this one step at a time."

- Hold up a paper strip and fold it in half to create a strip that's 2 x 5 inches. Open the strip and say, "I can see there are 2 parts to this strip now because the fold shows 2 pieces exactly the same size. I'm going to label each part of this strip as a fraction. Let's see. There

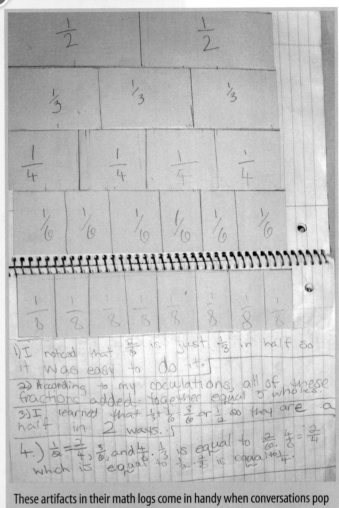

These artifacts in their math logs come in handy when conversations pop up about the relative size of fractions.

are 2 parts, so 2 is the denominator. This is 1 part, so I'll write '$\frac{1}{2}$' on this part and '$\frac{1}{2}$' on the other part."

- After students watch you fold and label the first strip of paper, instruct them to fold and label their strips. Once everyone has finished, probe further. Say, "Do you see that $\frac{1}{2} + \frac{1}{2} = 1$ whole?" It's very important to always refer to the parts and the whole.

- Pass out another strip of paper to each student. Say, "We'll fold this next strip into fourths." Repeat the steps for creating halves, and then demonstrate how to fold each of the halves in half to make fourths. Help children who may need assistance folding. Once everyone has folded their strips, have them discuss and label the fourths. Remind the students that $\frac{1}{4} + \frac{1}{4} + \frac{1}{4} + \frac{1}{4} = 1$ whole.

- Give each child a third strip of paper. Follow the steps above to show them how to fold the strip into fourths, and then show them how to fold each fourth in half again to create eighths. After the students fold and label their papers into eighths, remind them that $\frac{1}{8} + \frac{1}{8} + \frac{1}{8} + \frac{1}{8} + \frac{1}{8} + \frac{1}{8} + \frac{1}{8} + \frac{1}{8} = 1$ whole.

- Next, ask children to glue their paper strips into their math logs. Demonstrate how to glue down the strips so that they all line up as shown in the photo. That way kids can see that all of the strips are the same size. This is important because it helps children understand that $\frac{1}{2}$, $\frac{2}{4}$, and $\frac{4}{8}$ are all equivalent.

The last 2 paper strips will be folded into thirds and sixths. Folding the paper strip into thirds is more difficult, so you may decide to use this part of the activity the next day.

- Model how to fold the paper into thirds. Show the children that folding into thirds is tricky. A bit of visualizing and trial and error may be needed. But perseverance isn't just for kids; it's for teachers too. Sometimes it's good for kids to see teachers work something out. Stress the importance of patience and persistence. Have students label each part "$\frac{1}{3}$" and discuss the addition of all thirds to equal 1 whole.

- To create a paper strip folded into sixths, first fold the strip into thirds, and then fold each of the thirds in half. Voilà! Sixths. Have students label each part "$\frac{1}{6}$" and ask kids to tell the number sentence for 1 whole. $\left(\frac{1}{6} + \frac{1}{6} + \frac{1}{6} + \frac{1}{6} + \frac{1}{6} + \frac{1}{6} = 1 \text{ whole}\right)$

Mr. Klee's Squares

 Paul Klee (Getting to Know the World's Greatest Artists) by Mike Venezia

Paul Klee was a modern artist who used vivid colors to paint beautiful array-like paintings. (Another Klee task on page 16 helps teach arrays in multiplication.) Show children samples of his work, or read the very kid-friendly book *Paul Klee (Getting to Know the World's Greatest Artists)* by Mike Venezia.

Children will need paint in a variety of colors and brushes. (Crayons or markers will also work just fine.) To help the process move a bit more smoothly, provide each student with a paper marked with a 2 x 3 grid.

Tell children to color or paint each square or rectangle in their grid a different color. When the artists have completed their work, engage them in some mathematical conversation. You might ask a question such as, "What is the fraction of red- and green-colored squares in your array?" Remember to circle back to the idea that each of the unit parts put together equals 1 whole. $\left(\frac{1}{6} + \frac{1}{6} + \frac{1}{6} + \frac{1}{6} + \frac{1}{6} + \frac{1}{6} = 1 \text{ whole}\right)$

Whole Group

3.NF.A.1

Math Practices
4 Model with Mathematics
6 Attend to Precision

Kids love Klee's work because they feel it's easy to replicate.

Write About It: Ask each student to write about his array of Mr. Klee's squares in his math journal in draft form so that you can check his work before it becomes part of a bulletin board display.

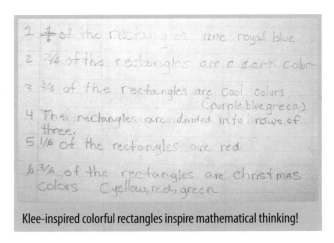

1. 1½ of the rectangles are royal blue

2. ⅔ of the rectangles are a dark color

3. ⅜ of the rectangles are cool colors (purple, blue, green)

4. The rectangles are divided into rows of three.

5. 1/6 of the rectangles are red

6. ⅜ of the rectangles are christmas colors (yellow, red, green)

Klee-inspired colorful rectangles inspire mathematical thinking!

3.NF.A.2 Understand a fraction as a number on the number line; represent fractions on a number line diagram.
3.NF.A.2a Represent a fraction $\frac{1}{b}$ on a number line diagram by defining the interval from 0 to 1 as the whole and partitioning it into b equal parts. Recognize that each part has size $\frac{1}{b}$ and that the endpoint of the part based at 0 locates the number $\frac{1}{b}$ on the number line.

Math Practices
6 Attend to Precision
7 Make Use of Structure

A Ribbon Ruler Tool

Your class can create a fractional number line using just a ribbon from the craft store (5 to 10 feet long and 2 inches wide) and a permanent marker.

- To begin, unroll the entire ribbon. Ask someone in the class to use the permanent marker to label one end of the ribbon with a "0." Ask another child to label the other end of the ribbon with a "1." You or a student will fold the ribbon in half so that the ends of the ribbon meet exactly.

- Ask, "Are you noticing that we're carefully folding the ribbon into equal sections? This is important because all parts of a fraction are equal." Press hard on the fold. Open the ribbon and label the pressed line "$\frac{1}{2}$."

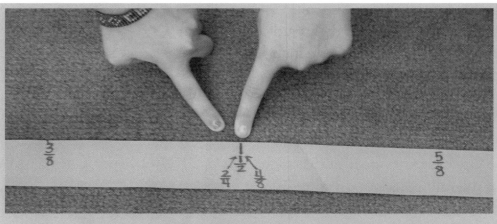

This easy-to-make ribbon ruler is a helpful tool that can be rolled up and unrolled throughout the year. You'll want to place this handy tool where your kids can easily access it!

- Fold the ribbon back in half and then carefully fold it in half again. Press hard and then unfold the ribbon. You, or a student, label the new presses "$\frac{1}{4}$" and "$\frac{3}{4}$." Under the $\frac{1}{2}$ mark, add the equivalent fraction, "$\frac{2}{4}$."

- Fold the ribbon in half again and then in quarters. Now fold it once more to create eighths. Press hard to create another fold mark, and then open the ribbon. You, or a student, mark the new fold lines "$\frac{1}{8}$," "$\frac{2}{8}$," "$\frac{3}{8}$," and so on up to "$\frac{8}{8}$."

- Be sure to point out that the $\frac{1}{2}$ fold is also equal to $\frac{2}{4}$ and $\frac{4}{8}$.

Extension: Let your students hop from fraction to fraction on the Ribbon Ruler. To play, say something such as, "Start at $\frac{1}{4}$. Hop another $\frac{1}{4}$. Where are you now?" or "Start at $\frac{3}{4}$. Hop backward $\frac{1}{4}$. Now where are you?" Kids love this game!

Jump Rope Fractions

Jump ropes from the dollar store or borrowed from your school's P.E. teacher make a great tool for this lesson. If jump ropes aren't available, ask the art teacher for long scrap ribbon.

Each group of 3 to 4 students will also need a handful of sticky notes (or index cards and tape or clothespins).

To begin, refer students back to the paper strips they glued into the math journals (Paper-Folded Fractions, page 83). Remind children how folding the paper strips into equal parts made it easy to both create and visualize equivalent fractions. Also remind them how important it is that the unit pieces of any whole are the same size.

Place students into small groups. Tell the children that they'll work together to divide a jump rope into fractional parts. Depending on your students' abilities or experience levels, you may decide to assign each group the same set of fractions, for example fourths $\left(\frac{1}{4}, \frac{1}{2}, \frac{3}{4}, \text{ and } 1\right)$, or you might give each group a different set of fractions to work on, such as thirds, sixths, and eighths.

Now it's time to pull out the jump ropes! Observe students as they work together to label the jump rope using the sticky notes. Give the kids time to talk this out and even to argue about it. Don't rush this process. Thinking is work. Your mathematicians are building strong and important concepts as they solve this problem. After students

Small Groups

3.NF.A.2b Represent a fraction $\frac{a}{b}$ on a number line diagram by marking off a lengths $\frac{1}{b}$ from 0. Recognize that the resulting interval has size $\frac{a}{b}$ and that its endpoint locates the number $\frac{a}{b}$ on the number line.

Math Practices
1 Solve Problems & Persevere
8 Express Regularity in Repeated Reasoning

Dividing a jump rope into equal fractional pieces is a challenging activity that requires using perseverance and problem-solving along with repeated reasoning.

have completed the task, ask groups to show their labels and explain their reasoning to their classmates.

Write About It: Ask students to explain how they showed the fractions with their jump ropes.

C ▶ P ▶ A

Whole Group

3.NF.A.3 Explain equivalence of fractions in special cases, and compare fractions by reasoning about their size.
3.NF.A.3a Understand two fractions as equivalent (equal) if they are the same size, or the same point on a number line.

Math Practices
4 Model with Mathematics
6 Attend to Precision
7 Make Use of Structure

Paper Plate Fractions

In this activity students compare the relative size of fractions using manipulatives made from simple paper plates.

To prepare, buy four 20-count packages of paper plates from a dollar store. You'll want each package to be a different color. Let's say you purchase yellow, red, blue, and purple paper plates. Use the yellow package of plates to represent the whole. Cut the red plates in half, the blue plates in fourths, and the purple plates in eighths. (We love to have our kiddos help us whenever possible, but for this activity we suggest that you cut the paper plates because it's important that the fraction pieces are precise.)

Make a set of paper plate fractions for each student. Store each set in a resealable plastic bag. (That way they'll be readily available to students throughout the year.) Each child will also need a fractional number line (a strip of paper on which an 8-inch-long line has been drawn and divided into 8 equal pieces).

- When you first give the paper plate fraction sets to the students, just let them explore. You'll be amazed at the things they'll discover.

- Listen to their conversations and encourage everyone to share their findings with their peers. Understanding equivalence is foundational for deep understanding of fractions, so don't rush their exploration of these manipulatives.

- After the students have had ample time to explore, pass out the fractional number lines and explain, "I'm going to place the actual paper plate pieces onto my number line, but you're going to draw pictures of these fractional parts on your number line."

- Quickly draw a "class-sized" number line divided into eighths on your whiteboard, or simply clip the paper plate fraction pieces to a jump rope as shown in the photo.

- Say, "First we must label our number line. How many sections does this line have? (8) Label the left hash mark with a '0' and the right hash mark with a '1.' All of the fractions that we model today will be between 0 and 1."

Jump ropes make the perfect number line and help students to visualize the relative size of fractions.

- Say, "Show me your yellow whole circle." Attach the actual yellow paper plate above the 1 on your number line model. Tell the students, "You are going to draw a circle on your number line to represent the 1 whole circle."

- Ask, "Which plate piece represents one half of the whole? (red) Can you find $\frac{1}{2}$ on your number line?" Ask, "How can you be sure this is the half?" (Students will have a variety of answers. Some may fold their number lines in half to prove that they are attending to precision.)

- Invite a student to come up to your number line and label the point on your line that represents one half. Instruct students to label the $\frac{1}{2}$-place on their number lines. Each student should also draw a half-circle above that spot.

- Continue this important conversation with the one-fourth and one-eighth pieces, following the same procedure used for halves. Ask volunteers to locate the correct hash marks as well as justify their answers.

- On another day ask students how else they could represent 1 whole. (2 halves, 4 fourths, and 8 eighths) Have them draw these equivalent fractions above the whole circle on their number lines.

C P A

Whole Group

3.NF.A.3b Recognize and generate simple equivalent fractions, e.g., $\frac{1}{2} = \frac{2}{4}$, $\frac{4}{6} = \frac{2}{3}$. Explain why the fractions are equivalent, e.g., by using a visual fraction model.

Math Practices
4 Model with Mathematics
6 Attend to Precision
7 Make Use of Structure

My Whole Hexagon

In this activity children explore equivalent fractions using pattern blocks as their concrete model. Next they're required to think pictorially as they trace the shapes. Finally, students must work at the abstract level as they label each of their drawings with the appropriate fraction.

Prepare a resealable plastic bag containing at least 2 yellow hexagons, 2 red trapezoids, 3 blue rhombi, and 6 green triangles for each student. Keep these bags and store them in a tub; you'll find yourself reaching for them time and time again. Each student will also need a sheet of paper.

- Say, "Each of you has a bag of pattern blocks and a sheet of paper. The yellow hexagon is going to represent your 1 whole for this activity. Can you find which block represents one half of the whole?" Walk around the room and observe your children.

- Choose a child to show his findings. Have the child prove that the red trapezoid is one half of the whole by stacking 2 red trapezoids on the yellow hexagon. Continue this same procedure for thirds (blue rhombi) and sixths (green triangles).

- Say, "Look at 1 red trapezoid. How could you make another trapezoid this same size using only 1 color of your blocks?" (Most children will quickly recognize that they can build a trapezoid using 3 green triangles.)

- Say, "Now you're going to trace your pattern blocks to document what you've discovered." Ask each student to trace the red

Mathematical Practices ask students to model with mathematics and attend to precision. This activity nails those practices!

trapezoid on her sheet of paper. Next have each student trace a second trapezoid right beside her first one. Say, "Using your triangles, divide 1 trapezoid into 3 equal parts."

• Continue by saying, "Now place an equal sign between the 2 drawings and write the fraction that each drawing represents under its picture." (Children should write "$\frac{1}{2}$" under the trapezoid and "$\frac{3}{6}$" under the trapezoid that was constructed with 3 green triangles.)

• Challenge your students. Say, "Find 1 blue rhombus in your bag. Is there another way to make a rhombus this same size with other blocks from your bag?" Walk around and observe. Ask a student who has shown that 2 green triangles equal 1 blue rhombus to explain his thinking to the class.

• Say, "Trace 1 blue rhombus on your paper. Trace a second blue rhombus next to the first one. Divide the second rhombus using the triangles. Place an equal sign between the 2 pictures. What fraction does each picture represent?" (Students should label the first picture "$\frac{1}{3}$" and the second "$\frac{2}{6}$.")

C ▶ P ▶ A
Whole Group

3.NF.A.3c Express whole numbers as fractions, and recognize fractions that are equivalent to whole numbers. *Examples: Express 3 in the form* $3 = \frac{3}{1}$*; recognize that* $\frac{6}{1} = 6$*; locate* $\frac{4}{4}$ *and 1 at the same point of a number line diagram.*

Math Practices
4 Model with Mathematics
6 Attend to Precision
7 Make Use of Structure

Stacking Wholes

In this activity students explore writing whole numbers as fractions. Each student needs a bag of pattern blocks (as prepared for My Whole Hexagon, page 90) and a sheet of paper.

- Begin by asking students to label the left side of their paper strip with a "0" and the right side with the number "1." Next ask students to take the yellow hexagon from their bag and place it above the number 1 on their number line.

- Ask students to find the red trapezoids in their pattern block bags. Say, "You're going to use the trapezoids to make a hexagon. How many trapezoids do you think it will take to make a whole hexagon?" (2 trapezoids are equivalent to 1 hexagon.)

- Say, "Since it takes 2 pieces to make 1 whole, what's each piece worth? (Each is worth $\frac{1}{2}$, and 2 halves make a whole.) Have the children place the trapezoids on their papers and write "$\frac{2}{2}$" above the 1 whole on their number lines. Say, "1 and $\frac{2}{2}$ are both worth 1 whole, so they have the same value on your number line."

- Next ask students to look for the blue rhombi. Ask students to find out how many rhombus pieces it will take to make a hexagon. (It takes 3 rhombus pieces to equal 1 hexagon.)

- Say, "Since 3 equal pieces make the whole, each blue rhombus piece is worth $\frac{1}{3}$." Have the students write "$\frac{3}{3}$" above the $\frac{2}{2}$ and 1 whole on their number line and then write the fractions next to the pattern blocks (see the photo).

- Ask the students to take out the green equilateral triangles. Say, "How many triangles does it take to make a whole hexagon? (6) Where will the $\frac{6}{6}$ go on our number line?" (Above the 1, $\frac{2}{2}$, and $\frac{3}{3}$.)

- Finally, have the students stack all of the pieces to create a "tower of wholes" as shown in the photo on the next page.

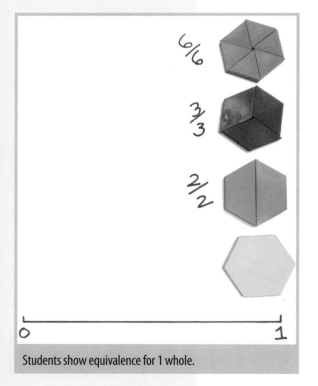

Students show equivalence for 1 whole.

Write About It:

Have the students pictorially represent each layer of the stack. Children love to color their representations and show the abstract fraction that names each layer.

Stacking and drawing the pattern blocks help children visualize fractional equivalency.

Fraction Fun

 Fraction Fun by David A. Adler

Adler's *Fraction Fun* is a delightful way to compare fractions and their relative sizes. The dog in this book is presented with 3 plates of pizza. Each plate has a different-sized slice. Will the dog choose $\frac{1}{2}$, $\frac{1}{4}$, or $\frac{1}{8}$? This story, along with the simple paper plate manipulatives, will get your students comparing fractions with ease. Start the activity by reading the first 13 pages of the book.

You'll need those paper plate fraction pieces you created in Paper Plate Fractions, page 88. These kid-friendly manipulatives will give your students the conceptual power to make a generalization. Generalizations and conjectures are what mathematical practice 8 is all about!

- Pass out the bags with the manipulatives. Tell students, "Take out 1 of each color except for the yellow one. Turn and talk to your neighbor about the size of each of the pieces. (Students should have $\frac{1}{2}$, $\frac{1}{4}$, and $\frac{1}{8}$.)

- Continue, "Let's pretend that the paper plate fractions are slices of pizza! Which piece would you want if you were very hungry?" Show the students the picture of the dog looking at the pieces of pizza in the story. "This dog is facing the same dilemma. Which

Whole Group

3.NF.A.3d Compare two fractions with the same numerator or the same denominator by reasoning about their size. Recognize that comparisons are valid only when the two fractions refer to the same whole. Record the results of comparisons with the symbols $>$, $=$, or $<$, and justify the conclusions, e.g., by using a visual fraction model.

Math Practices
4 Model with Mathematics
6 Attend to Precision
7 Make Use of Structure
8 Express Regularity in Repeated Reasoning

Students must attend to precision as they accurately draw and label the $\frac{1}{2}$, $\frac{1}{4}$, and $\frac{1}{8}$ pieces of pizza.

piece do you think he wants and why?" (Most children will want the $\frac{1}{2}$ piece because it's the biggest.)

- Ask, "How can $\frac{1}{2}$ be the biggest piece when it has a 2 in the denominator? I thought 2 was a small number!" Force your students to reason about the size of the denominator and how it relates to the size of the plate.

- Push further. Ask, "Why is the one-eighth slice of pizza so small? It has a very large number in the denominator!" Spend as much time as it takes for students to see this and understand.

Write About It: Tell each student to draw a picture of a pet looking at the 3 plates of pizza. Instruct them to draw the pizza pieces from largest to smallest. Next have students answer this question: "How can you tell which fraction is the largest if all the numerators are the same?"

GRADE 4

Cluster 4.NF.A Extend understanding of fraction equivalence and ordering.

It's a Piece of Cake

Whole Group

There's a bit of prep time involved for cutting out the pieces used in this activity, but it's your students who do the cutting. The understanding they gain as they prepare their pieces makes the time invested well worth it. It's a win-win situation!

Each child needs scissors and 4 congruent pieces of rectangular construction paper (at least 8 x 8 inches) in 4 different colors. Let's say the colors are red, green, blue, and orange.

- First and most important, you'll want to prove that these papers are congruent. Begin by saying, "You'll be folding and cutting 3 of your papers, so it's important that you understand and remember that each piece of paper is the same size."

- Take the red paper and say, "This will be the baker's cake pan. It's your whole. Let's label it 1 WHOLE." Give students time to label their papers.

- Remind students again that the other sheets of paper (which will represent the cakes) are all congruent to their red baker's pans. Say, "Once you fold and cut the other sheets of paper, they'll still fit into and fill up your baker's pan."

- Take the green paper and say, "Let's fold and cut this piece of paper in half vertically." Demonstrate a fold and cut. Say, "Write '$\frac{1}{2}$' on each piece" Place the 2 halves on the red whole and comment, "This cake is cut in half vertically. See, 2 halves equal 1 whole."

- Next have the students follow along as you demonstrate how to fold the blue paper in half vertically and then in half horizontally to create fourths. Open the paper and cut the pieces. Ask,

4.NF.A.1 Explain why a fraction $\frac{a}{b}$ is equivalent to a fraction $\frac{(n \times a)}{(n \times b)}$ by using visual fraction models, with attention to how the number and size of the parts differ even though the two fractions themselves are the same size. Use this principle to recognize and generate equivalent fractions.

Math Practices
4 Model with Mathematics
6 Attend to Precision

Understanding fractions is a piece of cake with this simple tool!

"How many pieces of blue cake do we have? How should we label the pieces? Yes, let's write '$\frac{1}{4}$' on each piece. Arrange your blue fourths on top of your green halves. What do you notice about the 4 fourths and the 2 halves?"

- Using the orange paper, have students repeat all the folds above, adding 1 more vertical fold so that when the paper is opened, there are 8 cake pieces. Ask, "What should we write on the orange pieces? Yes, write '$\frac{1}{8}$' on each piece. Cut the pieces. Place your orange eighths on top of your blue fourths. What do you notice about the 8 eighths and the 4 fourths?" (You want them to notice that they're congruent.)

These paper pieces make great "props" for all sorts of storytelling. As you know, CCSSM places emphasis on word problems and fractions. You can tell a story while the kids move the paper pieces to match your words. Have the class place their cut-outs onto their red baker's pans as you tell the following story:

Mrs. Lasater left this delicious cake in the teachers' lounge and Mrs. Kuhns ate $\frac{1}{4}$ of the cake. Next Mrs. Cliche ate $\frac{1}{8}$ of the cake, and then Mr. Murphy ate $\frac{1}{8}$ of the cake. How much cake was eaten? How much cake was left?

➡ $\frac{1}{2}$ of the cake was eaten and $\frac{1}{2}$ of it is left. $\left(\frac{1}{4} + \frac{1}{8} + \frac{1}{8} = \frac{1}{2}\right)$

Ask, "Do you notice that two $\frac{1}{8}$ pieces are the same size as the $\frac{1}{4}$ piece and two $\frac{1}{4}$ pieces are the same size as a $\frac{1}{2}$ piece?" It's important for students to use the paper manipulatives to recognize and generate equivalent fractions.

After you tell a few stories for them to solve, students will be eager to tell their own fraction stories for others to enjoy. Best of all, you'll see them switch two $\frac{1}{4}$ pieces for a $\frac{1}{2}$ piece and two $\frac{1}{8}$ pieces for a $\frac{1}{4}$ piece without you needing to say a thing. They'll figure this out. If they don't, be sure to point out how often there will be 2 fractional pieces that equal 1 other fractional piece.

Fraction Teams

Students create extra-large and extra-fun fraction number lines in this lesson. Be sure to keep these number lines handy for future fraction practice. They're great "go-to" tools!

Divide the students in your class into 4 groups. Provide each group with a roll of paper adding-machine tape. Each tape will need to be the same length. (If tapes are at least 8 feet long, students can easily fold them and make discoveries.) Before class you'll want to prepare a tape that's folded once and labeled "$\frac{1}{2}$."

- Begin by saying, "You're going to label your roll of paper with fractions. You might want to use a pencil first. Once you're sure you've labeled it correctly, you can go over your work in marker."

- Ask one group to label its tape in fourths, a second team to label its tape in eighths, the third team to label its tape in tenths, and the last group to label its tape in twelfths. Say, "This activity requires teamwork and perseverance. You'll have to listen to one another and reason things out."

- Expect chatter as they problem-solve. Restrain yourself from answering their questions. Let your students problem-solve. Should they measure? Fold? Estimate? Use the 1-foot-square tiles on the floor?

- Once everyone has completed their task, ask groups to explain how they found the fractions. Line up the tape you prepared as well as the students' tapes in rows so that children can see that the fractions equivalent to $\frac{1}{2}$ $\left(\frac{2}{4}, \frac{4}{8}, \frac{5}{10}, \text{ and } \frac{6}{12}\right)$ are in line. Ask, "What other fractions are at the same point of the line?" Seeing the position of a fraction on a number line is very helpful for students.

4.NF.A.2 Compare two fractions with different numerators and different denominators, e.g., by creating common denominators or numerators, or by comparing to a benchmark fraction such as $\frac{1}{2}$. Recognize that comparisons are valid only when the two fractions refer to the same whole. Record the results of comparisons with symbols >, =, or <, and justify the conclusions, e.g., by using a visual fraction model.

Math Practices
1 Solve Problems & Persevere
3 Construct Arguments & Critique Reasoning

Wow! When students line up their fraction number lines they easily see that all of the halves line up!

C **P** A

***Whole Group,
Small Groups***

4.NF.B.3 Understand a
fraction $\frac{a}{b}$ with $a > 1$ as a sum of
fractions $\frac{1}{b}$.
4.NF.B.3a Understand
addition and subtraction of
fractions as joining and
separating parts referring to the
same whole.

Math Practices
2 Reason Abstractly &
Quantitatively
4 Model with Mathematics
7 Make Use of Structure

GRADE ④

Cluster 4NF.B. Build fractions from unit fractions.

Another Serving of Cake, Please

It's time for students to pull out the baker's pans and cake pieces that
they created in It's a Piece of Cake, page 95. This time the focus is on
adding and subtracting the fractional pieces.

Be sure you've provided your students with plenty of practice
manipulating and exchanging the cake pieces. They should under-
stand that two $\frac{1}{8}$ pieces are equal to a $\frac{1}{4}$ piece and that two $\frac{1}{4}$ pieces are
equal to a $\frac{1}{2}$ piece. This knowledge will help them move more easily
into these addition and subtraction stories.

Begin with addition stories.

- Instruct students to manipulate their paper pieces as you tell this
  story. "The cake pan was empty when the baker placed 2 fourths of
  the cake and then one half of the cake into the pan. How much
  cake is now in the pan?" (1 whole cake) "Who can tell me the
  number model that matches this story problem?" $\left(\frac{1}{4} + \frac{1}{4} + \frac{1}{2} = 1\right)$

- Tell several similar addition stories, and then ask your students to
  tell their own stories to the class or to share stories in small groups.
  Notice this standard is not asking children to write, but rather to
  understand.

Now it's time to serve up some subtraction problems. You'll need
to model a few problems before your students can do this on their
own.

- Say, "Friends, let's start with a whole cake in the whole cake pan. It
  doesn't matter which cake pieces you use to fill up your cake pan,
  but as you're working to solve problems, you may find it helpful to
  exchange some of your pieces for equivalent ones. For example,
  you might decide to swap a one-half piece for 2 one-fourth pieces."

- Continue, "Ready? I had a whole chocolate cake. I served one
  fourth to Marilyn, and then one fourth to James. How much cake
  was left? Are we adding or subtracting? Who can tell the number
  model and the answer to this?" (Some may see this as 2 number
  sentences, "1 whole minus $\frac{1}{4}$ is $\frac{3}{4}$ and $\frac{3}{4}$ minus $\frac{1}{4}$ is $\frac{2}{4}$ or $\frac{1}{2}$!" Others
  may say, "1 whole minus $\frac{1}{2}$ is $\frac{1}{2}$.")

Mareo has a birthday cake. Colby ate one ½ Alex ate ⅜. How many pieces are left.

When children can choose their own topic, it's always more interesting.

- After a few more modeled subtraction stories, ask your students to take the lead and create problems that either the whole class or a small group can solve.

Write About It: Have each student write 3 or 4 addition and/or subtraction stories in his journal and circle the 1 problem he considers his best. Circled problems can be taken to the publishing stage and become material for a bulletin board or class book.

Like Denominator Addition

Here's a highly effective way to demonstrate how to add fractions with like denominators. This isn't necessarily the most efficient method for adding fractions, but it certainly drives the point home. You'll want to teach your students the more conventional method for adding fractions once they understand this concept.

    Always begin teaching this method using unit fractions. (A unit fraction always has a 1 as the numerator.)

- To begin, write the problem "$\frac{1}{3} + \frac{1}{3} =$" on the board. Say, "I can draw a picture to help me solve this problem." Draw a rectangle on the board. Say, "Look at the first fraction. There's a 3 in the

C P A

Whole Group

4.NF.B.3a

Math Practices
1 Solve Problems & Persevere
5 Use Tools Strategically

99

denominator, so I'll draw 2 vertical lines dividing the rectangle into thirds."

- Continue by saying, "Look at the numerator. It's 1, so I must color in 1 of the thirds." Use a colored pencil to shade in your drawing.

- Say, "Look at the denominator in the other fraction. It's 3. Since the rectangle I drew is already divided into thirds, I can continue. Let's look at the numerator in the next fraction. It's 1, so I must shade one third of the rectangle that's not already colored." Use a different colored pencil so students can see the two fractions.

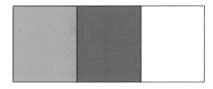

- Ask, "How many sections is this rectangle divided into? Yes, it's 3, so that represents the denominator. How many sections are colored? Yes, 2 sections are colored, so 2 is the numerator. $\frac{1}{3} + \frac{1}{3} = \frac{2}{3}$."

If you stick with unit fractions until your students' understanding is solid, they'll quickly grasp this method of adding fractions. Some students may be ready to move beyond addition of unit fractions (a skill that's above their grade level). To demonstrate how to add fractions with different denominators, see Unlike Denominator Addition, page 119.

What's the Whole?

Whole Group

The word "decompose" shows up a great deal in CCSSM. Children who can decompose a number truly understand the value of that number. Remember, decomposing means to be able to pull apart a number into 2 or more parts.

For students to decompose a fraction meaningfully, and not just by rote, they need plenty of experience actually decomposing fractions. Pattern blocks are the ideal tool for teaching this standard. By covering a larger pattern block with smaller pattern block pieces, it's easy to demonstrate decomposition.

Pull out the pattern blocks (with the exception of the orange squares and the thin tan rhombi). Ask students to follow along and copy as you demonstrate. Say, "If the trapezoid has a value of 1, and we can cover that trapezoid with 3 triangles, what fraction do the triangles represent? That's correct, $\frac{1}{3}$ and $\frac{1}{3}$ and $\frac{1}{3}$. So we can say $\frac{1}{3} + \frac{1}{3} + \frac{1}{3} = 1$."

Continue, "If the trapezoid has a value of 1 and we cover up the trapezoid with a triangle and a rhombus, what fractions do the triangle and rhombus represent? That's right, $\frac{1}{3}$ and $\frac{2}{3}$! So we can say $\frac{1}{3} + \frac{2}{3} = 1$."

Help students make more discoveries. Ask, "What if the hexagon has a value of 1 and we cover the hexagon with rhombi?" or "What If we cover the hexagon with triangles?" Encourage conversation. Ask the students to follow your lead and come up with questions to ask the class.

Write About It: After plenty of concrete experiences, ask students to record with drawings and fractional number sentences what they discovered about decomposing fractions.

Extension: Using colored tiles, have children create rectangles or rectilinear figures. Have students trace their tiles and write fractional number sentences to describe them. For example, "My rectangle has 10 squares, $\frac{3}{10}$ are red squares and $\frac{7}{10}$ are blue squares."

4.NF.B.3b Decompose a fraction into a sum of fractions with the same denominator in more than one way, recording each decomposition by an equation. Justify decompositions, e.g., by using a visual fraction model. *Examples:* $\frac{3}{8} = \frac{1}{8} + \frac{1}{8} + \frac{1}{8}$; $\frac{3}{8} = \frac{1}{8} + \frac{2}{8}$; $2\frac{1}{8} = 1 + 1 + \frac{1}{8} = \frac{8}{8} + \frac{8}{8} + \frac{1}{8}$.

Math Practices
4 Model with Mathematics
6 Attend to Precision

Pattern blocks help students to decompose fractions from a whole into smaller bits, making it easier to add fractions.

Whole Group

4.NF.B.3c Add and subtract mixed numbers with like denominators, e.g., by replacing each mixed number with an equivalent fraction, and/or by using properties of operations and the relationship between addition and subtraction.

Math Practices

3 Construct Arguments & Critique Reasoning
6 Attend to Precision
7 Make Use of Structure

Fraction Rulers

Fraction Rulers make it easy to compose and decompose fractions— they're handy for adding and subtracting fractions and mixed numbers too!

Each student needs six $2\frac{1}{2}$ x 12-inch paper strips, each a different color. (Let's say you use yellow, red, green, blue, pink, and orange.) Each student will also need two $2\frac{1}{2}$ x 12-inch white paper strips, an inch ruler, a pencil, and a gallon-sized resealable plastic bag.

Hooray! It's ruler construction day! More than just constructing rulers, you're helping students to construct knowledge.

- Pass out the plastic bags. Say, "Write your name and 'Fraction Rulers' on your bag." Tell your Einsteins that they also must put their names on all of the paper Fraction Ruler pieces they make today. This is nonnegotiable.

- Hand out 2 white paper strips to each student. Say, "These are your first 2 Fraction Rulers. These white papers are your wholes. Write 'whole' on both pieces." Point out that each paper strip is exactly 12 inches long. Ask kids to verify this by checking with their rulers.

- Next pass out the yellow paper strips. Say, "Hold your paper horizontally and fold it in half vertically." Ask, "What is half of 12 inches? How can you verify that your fold is accurate?" (Students may use a ruler to show that each piece equals 6 inches.)

- Continue by saying, "Label both halves of the strip with your name, and then turn the strip over and label both halves with the fraction '$\frac{1}{2}$.' Now carefully cut on the fold. These 2 pieces make your one-half Fraction Ruler."

- Pass out the red paper strips. Say, "You're going to make this strip into your one-thirds Fraction Ruler, so where do you think you'll need to make the folds and the cuts? Yes, you need to cut your paper into 3 equal pieces."

- Say, "Folding into thirds is tough, so we're going to use rulers. Match the 0 on the ruler with the edge of your paper. Make a mark at 4 inches and make a mark at 8 inches. Now carefully cut the paper on those marks. Write your name and the fraction '$\frac{1}{3}$' on each piece."

- Pass out the green paper strips. Say, "This is going to be your one-fourth Fraction Ruler. Let's think about the best way to measure and cut this paper into fourths."

- Your students may suggest folding the piece in half and then folding it again. They may suggest using a ruler and marking the paper off at 3, 6, and 9 inches. If you get two mathematically sound answers, pat yourself on the back because that means you're encouraging different thinking!

- Have students carefully cut the green strips into fourths. Write your name and the fraction '$\frac{1}{4}$' on each piece."

- Pass out the blue paper strips. Say, "This will be your one-sixth Fraction Ruler." Encourage conversation about the best way to measure and cut the 12-inch-long paper strip into sixths. Give students time to cut and label their pieces.

- Pass out the pink paper strips. Ask your class to discuss viable strategies for measuring and cutting the pink paper into 8 equal pieces. (Fold 3 times or measure $1\frac{1}{2}$ inches each.) Have students cut the paper and label the pieces into eighths.

- Finally, pass out the orange paper. Elicit strategies for the best and most precise way to divide the paper into twelfths. Give students time to cut and label their pieces.

- Ask kids to line up the Fraction Rulers and discuss how 2 or 3 of some Fraction Ruler pieces may equal 1 piece of a different Fraction Ruler. Encourage this exploration. Priceless conversation will be going on.

- To wrap things up, have kids place the Fraction Rulers in the resealable bags. Place the bags in a secure spot—but not so secure that kids can't readily put their hands on them!

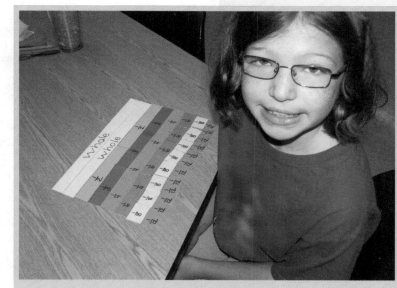

Fraction Rulers are great tools for partner or independent use.

Small Groups, Pairs

4.NF.B.3c

Math Practices
4 Model with Mathematics
7 Make Use of Structure

Let the Games Begin!

These math games provide a kid-friendly way to practice adding and subtracting mixed numbers. They're ideal for small group or partner work.

Players need a bag of Fraction Rulers, page 102, and a teacher-made fraction die.

To make a die, mark a wooden block or plastic cube with the numbers "$\frac{1}{2}$," "$\frac{1}{4}$," "$\frac{1}{8}$," "$\frac{1}{8}$," "$\frac{1}{12}$," and "$\frac{1}{12}$." To give students even more experience working with a range of fractions, make a second die. Mark 2 sides of that die "$\frac{1}{3}$," 2 sides of it "$\frac{1}{6}$," and 2 sides of it "$\frac{1}{12}$." Vary the dice the students use as they play the following games.

Exactly 1: The object of this game is to be the first player to fill 1 Whole Fraction Ruler exactly. In turn, each player rolls the die and covers her 1 Whole Fraction Ruler with smaller fraction pieces that match the number rolled on the die.

A roll that would make the total greater than 1 can't be used. For example, if a player has $\frac{3}{4}$ of her 1 Whole Fraction Ruler covered and rolls $\frac{1}{2}$, the total would be more than 1, so she must pass the die to her oponent. After each roll a player must say the addition sentence and the new total. ("I had $\frac{1}{4}$ and I rolled $\frac{1}{2}$; $\frac{1}{4}$ + $\frac{1}{2}$ equals $\frac{3}{4}$.") Whenever possible, players must exchange smaller Fraction Ruler pieces for larger ones.

Exactly 2: After students feel comfortable adding up pieces that equal 1 Whole Fraction Ruler, it's time for players to use both of their Whole Fraction Rulers! This game is played in the same manner as Exactly 1,

These kids are playing Exactly 2—and having fun adding mixed numbers.

except that the object is to be the first player to fill *both* Whole Fraction Rulers exactly.

Exactly 1 to 0: The object of this game is to be the first player to completely uncover 1 Whole Fraction Ruler and land exactly on 0. Each player sets up 1 Whole Fraction Ruler on his desk and covers it with smaller Fraction Ruler pieces. (Players may use any combination of Fraction Ruler pieces as long as the pieces cover the Whole Fraction Ruler exactly.)

In turn, each player rolls the die and removes the fractional amount shown on the die from his Whole Fraction Ruler. If a roll would result in a total that's less than 0, the player must pass. For example, if the player had $\frac{1}{8}$ left and rolled $\frac{1}{4}$, the player would pass the die to his opponent.

Exactly 2 to 0: After your charges have experience subtracting from 1 Whole Fraction Ruler, it's time to pull out 2 Whole Fraction rulers! This game is played in the same manner as Exactly 1 to 0, except that the object is to be the first player to uncover both Whole Fraction Rulers exactly.

Fraction Stories & Rulers

Got 5 or 10 minutes? Want a powerful way to spend that time? Give your students short doses of word problems. Two or three great problems a day will build their skills.

Each student will need his own set of Fraction Rulers, page 102.

Warm up your students by asking them questions that give them plenty of experience exchanging Fraction Ruler pieces for equivalent pieces.

- Say, "Place the 1 Whole Fraction Ruler on your desk. Place a one-fourth Fraction Ruler piece on top of it. Now add another one-fourth piece. What's the total? $\left(\frac{2}{4}\right)$ Can you exchange 2 one-fourth Fraction Ruler pieces for another Fraction Ruler piece?" $\left(\frac{1}{2}\right)$

Once your students are good and ready, ask them to solve addition problems.

- Say, "There was an empty bowl and Angela poured $\frac{1}{8}$ of a gallon of spaghetti sauce into the bowl. Gia poured $\frac{3}{8}$ of a gallon of spaghetti

Whole Group, Small Groups

4.NF.B.3d Solve word problems involving addition and subtraction of fractions referring to the same whole and having like denominators, e.g., by using visual fraction models and equations to represent the problem.

Math Practices
1 Solve Problems & Persevere
4 Model with Mathematics
7 Make Use of Structure

The queen had 6 jewels. $\frac{2}{6}$ of the jewels were sapphires. $\frac{3}{6}$ of the jewels were emeralds. The rest were rubies. How many were rubies?

$\frac{2}{6} + \frac{3}{6} = \frac{5}{6}$

$\boxed{\frac{1}{6}} + \frac{5}{6} = \frac{6}{6}$

There was 1 ruby.

This problem sparkles with fraction addition and subtraction.

sauce into the same bowl. How much spaghetti sauce was in that bowl?" $\left(\frac{4}{8} \text{ of a gallon}\right)$

- Provide many more stories that use fractions with the same denominators. Give problems with answers that can sometimes be exchanged for equivalent fractions. Limit the problems to answers that equal 1 whole or less. Use their names and their interests, and you'll have your students' attention!

When your class is ready for subtraction story problems, instruct them to cover their 1 Whole Fraction Rulers with the smaller pieces from a different Fraction Ruler set. Again, stick to story situations with like denominators.

- Say, "Cover 1 Whole Fraction Ruler with all of the one-fourth Fraction Ruler pieces. Now take away 3 fourths. What's left?" $\left(\frac{1}{4}\right)$

- Ask, "Can anyone think of a story that would match the number sentence 1 take away $\frac{3}{4}$?" If they can't, you should provide a simple story such as, "I had 1 salad and I served $\frac{3}{4}$ of the salad. How much salad was left?" $\left(\frac{1}{4}\right)$

- Provide your students with more subtraction stories that use like denominators. Say, "The baker had 1 bag of flour. He used $\frac{2}{3}$ for pies. How much flour is left?" Ask students, "What Fraction Ruler pieces are you going to use to show this problem on your 1 Whole Fraction Ruler? Yes, the thirds. So place 3 thirds on your 1 Whole Fraction Ruler and then take away 2 of the thirds." $\left(\frac{1}{3}\right)$

It's important that your students show their answers with the Fraction Ruler pieces. This concrete, hands-on tool helps them to develop a strong and flexible understanding of equivalency.

Once your students are ready, ask them to listen to one another's addition and subtraction stories. Give them time to manipulate and exchange their Fraction Ruler pieces and insist that they show the correct answers with their rulers.

Write About It: Ask students to write addition and/or subtraction fraction number sentences using like denominators. Have students check their work using the fraction rulers.

Multiply with Pattern Blocks

This lesson uses pattern blocks to help students visualize what happens when a whole number is multiplied by a fraction.

To begin, remind students that the yellow hexagon represents 1 whole. Review the fractional values for the other pattern blocks in relation to the yellow hexagon. The red trapezoid is $\frac{1}{2}$ because it takes 2 trapezoids to cover the hexagon; the blue rhombus is $\frac{1}{3}$ because it takes 3 rhombi to cover the hexagon; and the green triangle is $\frac{1}{6}$ because it takes 6 triangles to cover 1 hexagon.

- Say, "Using the pattern blocks as a visual model, show me 6 groups of $\frac{1}{6}$." The students should model this by showing 6 of the green triangles since they are each $\frac{1}{6}$ of the whole.

- Continue by saying, "Now let's think about the multiplication problem $6 \times \frac{1}{6}$. How many wholes (hexagons) can you make with the 6 green triangles? That's right, 1, so 6 times $\frac{1}{6}$ equals $\frac{6}{6}$ or 1 whole."

Whole Group, Small Groups

4.NF.B.4 Apply and extend previous understandings of multiplication to multiply a fraction by a whole number.
4.NF.B.4a Understand a fraction $\frac{a}{b}$ as a multiple of $\frac{1}{b}$. *For example, use a visual fraction model to represent $\frac{5}{4}$ as the product $5 \times \left(\frac{1}{4}\right)$, recording the conclusion by the equation $\frac{5}{4} = 5 \times \left(\frac{1}{4}\right)$.*

Math Practices
4 Model with Mathematics
5 Use Tools Strategically

Pattern blocks help build fractional understanding.

- Ask, "Ready for another equation? Use the pattern blocks to model 4 groups of $\frac{1}{3}$." (Students should show 4 blue rhombi since each rhombus piece is $\frac{1}{3}$ of the whole.)

- Ask, "How many wholes (hexagons) can you make with 4 blue rhombi or 4 thirds?" Help students see that they can completely cover the hexagon with 3 of the rhombi but they'll have 1 rhombus left over. Pattern blocks make it easy to see that $4 \times \frac{1}{3} = \frac{4}{3}$ and that $\frac{4}{3} = 1\frac{1}{3}$.

- Once students have had plenty of hands-on experiences exploring the multiplication of unit fractions using pattern blocks, let them take this concept to the pictorial level. Their results will look similar to the awesome illustrations you see pictured here!

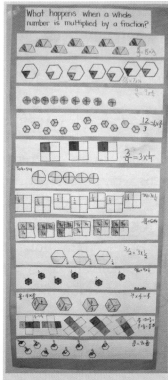

The student drawings create a very visually pleasing anchor chart.

C P A

*Whole Group,
Small Groups, Pairs*

4.NF.B.4b Understand a multiple of $\frac{a}{b}$ as a multiple of $\frac{1}{b}$, and use this understanding to multiply a fraction by a whole number. *For example, use a visual fraction model to express* $3 \times \left(\frac{2}{5}\right)$ *as* $6 \times \left(\frac{1}{5}\right)$, *recognizing this product as* $\frac{6}{5}$. *(In general,* $n \times \left(\frac{a}{b}\right) = \frac{(n \times a)}{b}$.*)*

Math Practices

3 Construct Arguments & Critique Reasoning

5 Use Tools Strategically

Pattern Blocks, Again!

In this activity students once again manipulate pattern blocks to gain a concrete understanding of what happens when a fraction is multiplied by a whole number.

To begin, pass out pattern blocks to each group of children and conduct a quick review of the fractional values for each of the pattern block pieces. (For a quick refresher, see Multiply with Pattern Blocks, page 107.)

- Say, "Work with the people in your group to show me 2 groups of two thirds." Students should pull out 2 sets of 2 rhombi. Affirm that everyone has 4 rhombi.

- Continue by asking, "What equation does this represent?" $\left(2 \times \frac{2}{3} = \frac{4}{3}\right)$ Give kids time to work this out.

- Ask, "Can you show me 4 groups of one third?" (Students should place 4 rhombi on their workspace.) Ask, "What equation does this represent?" $\left(4 \times \frac{1}{3} = \frac{4}{3}\right)$

- Ask, "How many rhombi did you use in $2 \times \frac{2}{3}$?" (4) "How many rhombi did you use in $4 \times \frac{1}{3}$?" (4)

- Now challenge your students. Ask, "How does the equation $4 \times \frac{1}{3} = \frac{4}{3}$ relate to the equation $2 \times \frac{2}{3}$?" (Both equations have the same answer. They both equal $\frac{4}{3}$ or $1\frac{1}{3}$.)

- Say, "Yes, both equations do have the same answer. Why do you think that's true? What's happening?" You want the kiddos to see that 4 groups of $\frac{1}{3}$ is the same as 2 groups of $\frac{2}{3}$. The rhombi will be a concrete model that proves this.

This is a deep concept. Students need repeated practice working in small groups or with a partner solving similar problems before they can develop a solid understanding of this concept and really "own" it. Have students take their learning to the pictorial level by asking them to record their findings as shown in the photo.

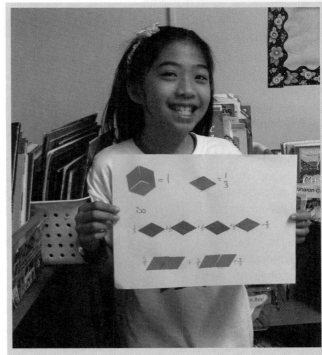

The pattern blocks make it easy to understand that 4 groups of $\frac{1}{3}$ are equivalent to 2 groups of $\frac{2}{3}$.

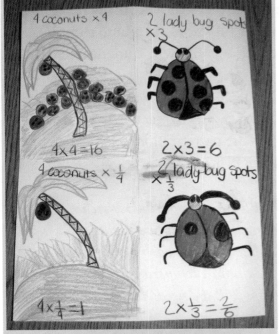

Kids love the nonsense in *Multiplying Menace, the Revenge of Rumpelstiltskin*.

Extension: Read *Multiplying Menace, the Revenge of Rumpelstiltskin*, by Pam Calvert to your class, and then ask students to illustrate a fractional multiplication story similar to the events in the kingdom where Rumpelstiltskin wreaked havoc.

C ▶ P ▶ A

*Whole Group,
Small Groups*

4.NF.B.4c Solve word problems involving multiplication of a fraction by a whole number, e.g., by using visual fraction models and equations to represent the problem. *For example, if each person at a party will eat $\frac{3}{8}$ of a pound of roast beef, and there will be 5 people at the party, how many pounds of roast beef will be needed? Between what two whole numbers does your answer lie?*

Math Practices

1 Solve Problems & Persevere
4 Model with Mathematics

Full House

 Full House: An Invitation to Fractions by Dayle Ann Dodds

This book inspires word problems that involve multiplying a fraction by a whole number, and that's what this standard is all about. As you read the story to your students, have your pattern blocks ready. (If a quick review of the fractional values for each of the pattern block pieces is necessary, see Multiply with Pattern Blocks, page 107.)

Read this book to your class once for sheer enjoyment. In the story, Miss Bloom owns the Strawberry Inn. It has 6 rooms for 5 guests and Miss Bloom. When you read the book for the second time, represent each guest and Miss Bloom with a green triangle pattern block piece.

Point out to students that when the whole inn is full you'll have 6 green triangles and that the 6 triangles form 1 whole hexagon. The pattern blocks provide students with a concrete way to visualize and understand fractions. Each triangle represents $\frac{1}{6}$ of the hexagon.

You'll also have the opportunity to recap equivalent fractions as you go along. When the second triangle is in place, ask the children if $\frac{2}{6}$ can be represented in another way. ($\frac{1}{3}$ can be shown with 1 rhombus or 2 triangles.) Do this again with $\frac{3}{6}$ to show equivalence to $\frac{1}{2}$ (the red trapezoid, or a blue rhombus and a triangle). Help students see that $\frac{4}{6}$ is equivalent to $\frac{2}{3}$.

Next, display the following problem, inspired by the characters in the story, and make pattern blocks available to students.

Miss Bloom is preparing ice cream with toppings for dessert. She needs $\frac{1}{6}$ of a cup of chocolate chips for each person and $\frac{2}{3}$ cup of ice cream for each person. How many cups of chips will she need? How many cups of ice cream?

➡ Miss Bloom will need 1 cup of chips and 4 cups of ice cream.

Pattern blocks help students see that Miss Bloom will need 4 cups of ice cream.

The pattern blocks are a concrete way of helping students see that 6 groups of $\frac{1}{6}$ (the green triangle) = $\frac{6}{6}$ or 1 whole (the yellow hexagon); and that 6 groups of $\frac{2}{3}$ (the blue rhombus) = $\frac{12}{3}$ or 4 (yellow hexagons).

Once your students have had lots of problems to "chew" on, invite your mathematical authors to create similar problems for classmates to solve.

Powerful Kid-Created Problems

Individuals

4.NF.B.4c

Math Practices
1 Solve Problems & Persevere
4 Model with Mathematics

Is this math or writing? Yes. The answer is yes. After your students have worked with concrete materials and understand what's happening when a fraction is multiplied by a whole number, it's time for story writing.

Tell the class, "I've told you lots of math story problems; now it's your turn to write a great problem where a fraction is multiplied by a whole number. I want you to be clever and use language that will make others want to read your problem. First you'll write a draft problem including the solution. I'll edit your story for mathematics, word choice, and conventions."

After problems have been edited by you for conventions and accuracy, pass out special paper for kids to write their own problem and math sentence, and to illustrate the solution.

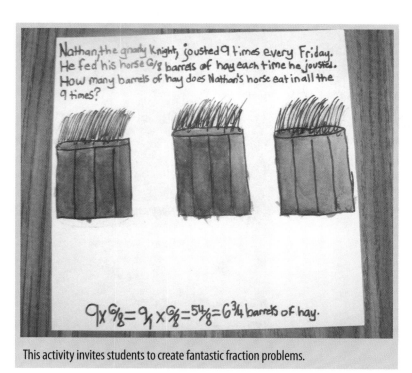

This activity invites students to create fantastic fraction problems.

Cluster 4.NF.C. Understand decimal notation for fractions, and compare decimal fractions.

C ▸ P ▸ A
Individuals

4.NF.C.5 Express a fraction with denominator 10 as an equivalent fraction with denominator 100, and use this technique to add two fractions with respective denominators 10 and 100. *For example, express $\frac{3}{10}$ as $\frac{30}{100}$, and add $\frac{3}{10} + \frac{4}{100} = \frac{34}{100}$*

Math Practices
6 Attend to Precision
7 Make Use of Structure

Fishy Bowls

You'll hook 'em with this fishy lesson! Each child will need a piece of white construction paper, scissors, colored pencils, and markers or watercolors.

To begin, each child draws and then cuts out a symmetrical fish bowl at least 8 inches wide and 8 inches high. Next each student draws 10 fabulous fish inside her fish bowl and colors the fish by adding spots, stripes, or other distinguishable traits. (To make this even more spectacular, allow kids to brush blue watercolor paint over the fish.)

Next, have each student write sentences about the combination of fish in her bowl using fractions. Children should first use 10 as the denominator and then move on to using 100 as the denominator. For example, "$\frac{6}{10}$ or $\frac{60}{100}$ of my fish have spots."

Write About It: Ask students to write addition sentences about their fish using fractions in their math journals. For example, "$\frac{20}{100}$ of my sea creatures have a shell and $\frac{80}{100}$ of my my sea creatures don't have a shell. That's $\frac{20}{100} + \frac{80}{100} = \frac{100}{100}$ or 1 whole!"

Underwater creatures bring this standard to picture form.

Loops & Loops

This task demonstrates the value of hundredths using proportionality. First students count 100 pieces of colored cereal. Next they sort that group of cereal pieces into smaller groups by color. Working together, students determine the fractional amount in each color group.

Place kids in small groups of 3 or 4. Provide each group with a cup of colored loop cereal. Each group also needs a plain sheet of paper.

- Instruct each group to work together to count out 100 pieces of cereal from their cup. Once all 100 pieces of cereal are counted, ask the children to sort the 100 cereal pieces by color as shown in the photo.

- Next have students record the total number of pieces of cereal in each color group on their paper. For example, red = 22 pieces. Point out that the total of all of the different color groups will always equal 100.

- Say, "Since there are 100 pieces of cereal, the denominator will always be 100. That means if you had 22 pieces of red cereal out of the total 100 pieces, the fraction of red pieces would be $\frac{22}{100}$."

- Once all records are complete, tell students to go back and add all of their numerators. Ask, "Do your numerators add up to 100? They should."

C P A
Small Groups

4.NF.C.6 Use decimal notation for fractions with denominators 10 or 100. *For example, rewrite 0.62 as $\frac{62}{100}$; describe a length as 0.62 meters; locate 0.62 on a number line diagram.*

Math Practices
6 Attend to Precision
7 Make Use of Structure

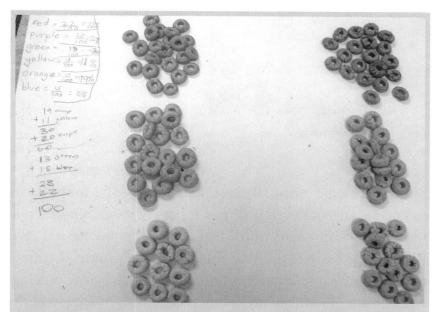

It's easiest for kids to count and sort the cereal if all of the pieces are spread out on a sheet of paper as shown here.

4.NF.C.6

Math Practices
6 Attend to Precision
7 Make Use of Structure

Roll to a Meter

The meter stick isn't just for measuring; it's a handy reference for decimals too!

Each pair of students will need 2 dice, a meter stick, 20 base-10 longs, and 20 base-10 ones. Each student will need a recording sheet similar to the one shown in the photo.

In this game children use the 10 centimeter "long" to represent the decimal 0.1 and the 1 centimeter "one" to represent the decimal 0.01. Because the longs measure 10 centimeters and the ones measure 1 centimeter, these manipulatives work beautifully alongside the meter stick.

The meter stick is placed between the 2 players. In turn, players roll the dice. The player who rolls must total the numbers on the dice. If the total is 7, that player places 7 centimeter cubes along his side of the meter stick. If the player's next roll is a 6 and a 5, he would add 11 more centimeter cubes (which would be 1 long and 1 "one") to his side of the meter stick, making it 18 centimeters or 0.18 meter. The first player to reach 1 meter wins the game!

As students record their moves on their worksheets, they're linking fractions to decimals.

Explain to your mathematicians that they're expected to exchange every 10 centimeter cubes for 1 long. Be sure to let kids know that with each roll they must say something such as, "I'm adding 11 centimeters or 0.11 of a meter. I now have 0.18 of a meter. I need 0.82 meter more to reach 1 meter." Some of your students will know to exchange 10 ones for a long before placing the "'ones" alongside the meter stick; however, some students will need to see 10 "ones" lined up before understanding that an exchange is necessary.

As an added benefit, students build benchmarks of the metric system while playing Roll to a Meter.

How Does Your Garden Grow?

Individuals

In this artful activity, students create a visual model of a garden. Along the way they get lots and lots of practice adding decimals!

Each child needs an index card and a 10 x 10 grid. You'll want to go online and find a few photos of botanical gardens (such as Monet's garden in Giverny) and/or photos of a farmer's field to share with students.

- Tell students that to solve this math task they'll get to think like landscape designers or agricultural planners! Begin by showing students the photos you collected. Point out how each crop of flowers or fruits/vegetables is planted together in groups.

- Pass out the 10 x 10 grids. Instruct students to divide their grids into small rectangles and/or rectilinear shapes. Explain that each 0.01 rectangle on the grid will be planted with a specific crop.

- Planning is critical. There must be at least 6 different types of plants represented in the garden, but no 2 types of crops may take up the same number of grids. For example, if 0.04 of the garden is spinach, no other plant may equal 0.04 of the whole garden.

4.NF.C.7 Compare two decimals to hundredths by reasoning about their size. Recognize that comparisons are valid only when the two decimals refer to the same whole. Record the results of comparisons with the symbols >, =, or <, and justify the conclusions, e.g., by using a visual model.

Math Practices
5 Use Tools Strategically
7 Make Use of Structure
8 Express Regularity in Repeated Reasoning

- Tell kids to color and label each section of their grid with a specific planting or crop. There may be no left-over (uncolored) rectangles.

- For some students this can be tricky and they end up doing a lot of erasing. So, you might consider giving your students one grid to use as a draft and another to use for the final copy.

- Once the students' visual models (their grids) are filled with plants, ask them to write number models on their index cards using the <, >, and/or = symbols to compare the amounts of the different crops in their gardens. For example, "My garden is 0.04 carrots and 0.07 pumpkins. 0.04 < 0.07."

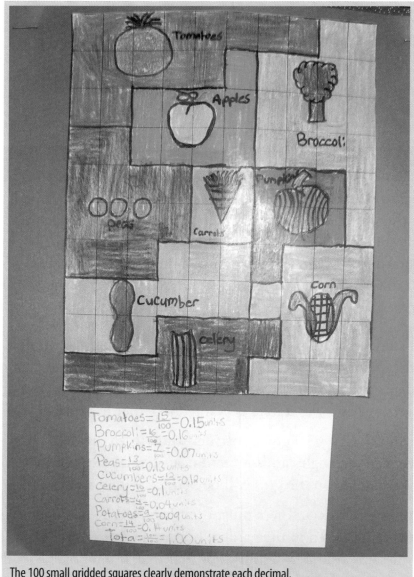

The 100 small gridded squares clearly demonstrate each decimal.

GRADE ⑤

Cluster 5.NF.A Use equivalent fractions as a strategy to add and subtract fractions.

Building Fractions with Box Turtles

 Picture Pie by Ed Emberly

As students build these delightful turtles (inspired by the creations from Mr. Emberly's *Picture Pie*), they deepen their understanding of equivalent fractions. Don't be surprised if you feel the urge to create a bulletin board inspired by this book—you're not alone. The images are charming!

Each child needs at least 4 paper circles. The circles shown here are about 4 inches in diameter, but the actual size doesn't matter. Make each circle a different color; for example, white, red, green, and yellow.

Once the circles are cut, leave the white ones whole. Cut the red ones in half, the green ones in fourths, and the yellow ones in eighths. Color coding makes it easy for students to see and compare the fractional size of each piece.

- After enjoying *Picture Pie* with your students, review the pages that show how the artist used a circle divided into pie shapes to create all of the pictures in the book.

- Pass out the paper fraction pieces and encourage kids to explore. Suggest they place 2 red halves on the 1 white whole. Ask, "Are there any other pieces that can cover, or equal another piece?"

- This provides so much content for a group discussion. For example, 2 green pieces equal 1 red piece, so $\frac{2}{4}$ is equivalent to $\frac{1}{2}$ and 4 yellow pieces equal 1 red piece, so $\frac{4}{8}$ is also equivalent to $\frac{1}{2}$. In order for students to develop a deep conceptual understanding of fractions, they must comprehend equivalency, so don't rush this exploration.

- After they've had time to explore, ask, "Can you use the fraction pieces to show me what $\frac{1}{4} + \frac{2}{8}$

Whole Group

5.NF.A.1 Add and subtract fractions with unlike denominators (including mixed numbers) by replacing given fractions with equivalent fractions in such a way as to produce an equivalent sum or difference of fractions with like denominators. *For example, $\frac{2}{3} + \frac{5}{4} = \frac{8}{12} + \frac{15}{12} = \frac{23}{12}$. (In general, $\frac{a}{b} + \frac{c}{d} = \frac{(ad + bc)}{bd}$.)*

Math Practices
2 Reason Abstractly & Quantitatively
6 Attend to Precision

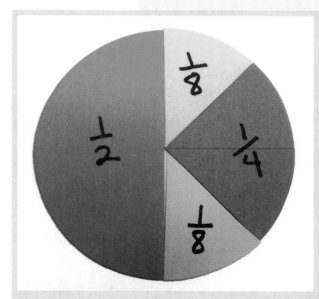

The fraction pieces make it easy to see that $\frac{1}{2} + \frac{1}{4} + \frac{2}{8} = 1$ whole.

117

equals?" It's important for students to understand that the answer to this problem can be represented in multiple ways: $\frac{4}{8}$, $\frac{2}{4}$, or $\frac{1}{2}$.

- Challenge students. Ask, "How many ways can you make 1 whole? Write an equation for each way." (They could have $\frac{1}{2} + \frac{2}{4} = 1$, $\frac{2}{8} + \frac{3}{4} = 1$, $\frac{1}{2} + \frac{1}{4} + \frac{2}{8} = 1$, or other solutions.) Use this opportunity as a springboard for mathematical conversations. The results could create a wonderful anchor chart for your classroom wall.

For some classrooms, this exploration of the paper fraction pieces might be enough for the day. That's fine. If not, press on. It's time to build turtle shells! Students will need tape or glue and access to pattern blocks.

> **✓ QUICK TIP**
>
> *See Like Denominator Addition, page XX for using this model to add fractions with like denominators.*

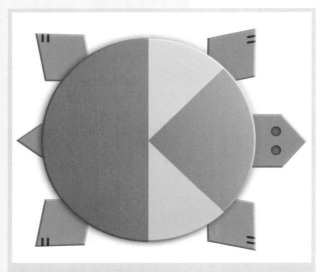

The turtle's polygon features will link to later lessons in geometry.

- Show students how to construct a turtle shell by covering 1 full circle with the colorful paper fraction pieces. Explain that they must use fractions with different denominators (for instance a turtle shell made of just 2 one half pieces isn't acceptable). Fraction pieces can be taped or glued to the white paper circle.

- To complete the turtle's features, let students trace pattern blocks. For example, the green triangle makes a tail, the blue rhombi make legs, and the orange square plus the green triangle could be used to form the pentagon head. Have students label the geometric pieces. (That way they're attending to precision and they're working on standards they'll explore later in geometry!)

- Once the turtles are complete, ask students to write the equation that matches the fractional pieces they used to create the shell on the back side of their turtle. Encourage children to talk to their friends about the fractions they've used.

- You'll want to save the students' turtles for future learning; they're used again in the activity on page 131.

Write About It: Have students write about and draw 2 or more ways they could build a turtle shell. Remind them to use fraction pieces with at least 2 different denominators for each shell.

Extension: Ask students how many turtles they could build using 3 one-eighth pieces, 5 one-fourth pieces, and 2 one-half pieces. Require them to support their answers with an equation, words, and drawings. (2 whole turtles can be built with $\frac{5}{8}$ left over.)

Unlike Denominator Addition

The technique presented here is very effective at helping students visualize the addition of fractions with unlike denominators. However, once your students understand this concept, you'll want to move them on to using the more time-efficient traditional method.

- To begin, write "$\frac{1}{3} + \frac{1}{4} =$" on the board. Say, "To show you how to add these fractions, I'll draw a rectangle. Now, look at the first fraction. The denominator is 3, so I must divide the rectangle into thirds." Draw 2 vertical lines to divide the rectangle into thirds. Say, "The numerator is 1, so I'll color in 1 of the thirds."

- Continue by saying, "Look at the second fraction. The denominator is 4, but my rectangle is divided into thirds. It must now be divided into fourths." Draw 3 lines horizontally, dividing the rectangle into fourths.

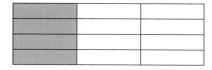

- Say, "Look at the fourths, there are 3 small units in each fourth. Since we're adding $\frac{1}{4}$, I'll color in 3 small units. I can't color in any units that have already been colored. Color in 3 units using a different color (this helps the children see this as 2 separate fractions being added).

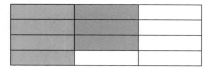

C P A

Whole Group

5.NF.A.2 Solve word problems involving addition and subtraction of fractions referring to the same whole, including cases of unlike denominators, e.g., by using visual fraction models or equations to represent the problem. Use benchmark fractions and number sense of fractions to estimate mentally and assess the reasonableness of answers. *For example, recognize an incorrect result $\frac{2}{5} + \frac{1}{2} = \frac{3}{7}$, by observing that $\frac{3}{7} < \frac{1}{2}$.*

Math Practices
1 Solve Problems & Persevere
5 Use Tools Strategically

- Say, "Now I count the small rectangles. There are 12, so that's the denominator. Next I count the small units that are colored. There are 7. So 7 is the numerator. The answer is $\frac{7}{12}$!"

Be sure to give your students the opportunity to work through many of these problems with a partner or on their own.

GRADE ⑤

Cluster 5.NF.B Apply and extend previous understandings of multiplication and division to multiply and divide fractions.

Ⓒ ▸ Ⓟ ▸ Ⓐ
Whole Group

5.NF.B.3 Interpret a fraction as division of the numerator by the denominator $\left(\frac{a}{b} = a \div b\right)$. Solve word problems involving division of whole numbers leading to answers in the form of fractions or mixed numbers, e.g., by using visual fraction models or equations to represent the problem. *For example, interpret $\frac{3}{4}$ as the result of dividing 3 by 4, noting that $\frac{3}{4}$ multiplied by 4 equals 3, and that when 3 wholes are shared equally among 4 people each person has a share of size $\frac{3}{4}$. If 9 people want to share a 50-pound sack of rice equally by weight, how many pounds of rice should each person get? Between what two whole numbers does your answer lie?*

Math Practices
3 Construct Arguments & Critique Reasoning
4 Model with Mathematics
5 Use Tools Strategically
6 Attend to Precision

Doorbell Rang

 The Doorbell Rang by Pat Hutchens

This humorous story about sharing chocolate chip cookies is a tried and true favorite. Here's a completely different way to look at those freshly baked cookies that'll help your charges understand the division of 1 whole.

You'll need 12 Unifix cubes (or other snap cubes) to demonstrate the sharing that goes on in the story.

Manipulatives help kids see that 12 cookies divided among 9 people gives each person $1\frac{1}{3}$ cookies.

- Say, "Let's think of that full plate of cookies as 1 whole. I'll use Unifix cubes to represent the 12 cookies." Snap together a stick of 12 cubes and say, "This stick represents 1 whole plate of cookies."

- Continue by saying, "12 cookies divided between 2 children would give each child 6 of the 12 cookies or $\frac{1}{2}$ of the cookies." Split the stick of Unifix cubes in half to illustrate this point.

- Invite a student to manipulate the Unifix cubes to show 12 cookies divided among 4 children. Agree, "Yes, each child would get 3 of the 12 cookies, which is equivalent to $\frac{3}{12}$ or $\frac{1}{4}$."

- Now it's time for a challenge. Say, "What if we had to find fractional parts of a cookie in order to give every child a fair share? For example, what if I had 12 cookies and I had to divide them among 9 people? How many cookies would each person get?" (Each person would get 1 and $\frac{1}{3}$ of a cookie.)

- Ask students to solve this problem by drawing an illustration or using other manipulatives. Remember to praise those selections! This is a great example of using tools strategically. Allow time for students to share and support their reasoning.

Write About It: Ask students to share 3 cookies with 4 people. Require them to show their answer on a number line.

When asked to share 3 cookies with 4 people, some kids will divide each cookie in quarters and give each person $\frac{1}{4}$ of each cookie. Others will see that $3 \div 4 = \frac{3}{4}$. This understanding takes time to develop. Be patient!

C ▶ P ▶ A

Whole Group

5.NF.B.4 Apply and extend previous understandings of multiplication to multiply a fraction or whole number by a fraction.

5.NF.B.4a Interpret the product $\left(\frac{a}{b}\right) \times q$ as a parts of a partition of q into b equal parts; equivalently, as the result of a sequence of operations $a \times q \div b$. For example, use a visual fraction model to show $\left(\frac{2}{3}\right) \times 4 = \frac{8}{3}$, and create a story context for this equation. Do the same with $\left(\frac{2}{3}\right) \times \left(\frac{4}{5}\right) = \frac{8}{15}$. (In general, $\left(\frac{a}{b}\right) \times \left(\frac{c}{d}\right) = \frac{ac}{bd}$.)

Math Practices

3 Construct Arguments & Critique Reasoning

4 Model with Mathematics

7 Make Use of Structure

8 Express Regularity in Repeated Reasoning

Thinking About Groups

Back in the third grade, students learned to visualize multiplication. For example, for "3 groups of 5" they might envision 3 hands with 5 fingers each or they might see 3 pentagons with 5 sides each. It's now time to extend that reasoning to fractions.

Each student needs a bag of assorted pattern blocks for this activity. Pattern blocks help students visually "decompose" a whole into smaller pieces, and that makes it easier to understand multiplying fractions.

- To begin, remind students that the yellow hexagon represents 1 whole. Review the fractional values for the other pattern blocks in relation to the yellow hexagon. (For a refresher, see Multiply with Pattern Blocks, page 107.)

- Proceed by asking, "Can you find a way to show me $3 \times \frac{1}{2}$, or in other words, 3 groups of one half?" Someone might show 3 red trapezoids to illustrate 3 groups of $\frac{1}{2}$. (This is because 2 trapezoids equal 1 hexagon.) If this happens, it's critical that the student identifies which piece is the whole (the hexagon) and explains how many wholes can be made with the 3 halves. Write the equation "$3 \times \frac{1}{2} = 1\frac{1}{2}$" on the board for all to see.

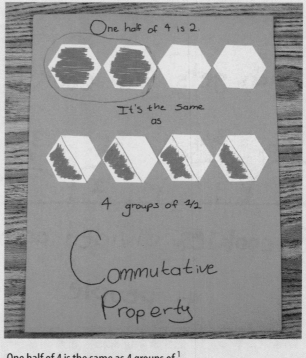

One half of 4 is the same as 4 groups of $\frac{1}{2}$.

- Continue this activity by asking kids to use the pattern blocks to solve similar problems. For example, ask, "Can you find a way to show me $4 \times \frac{1}{3}$, or 4 groups of one third and can you find a way to show me $6 \times \frac{1}{6}$, or 6 groups of one sixth." ($4 \times \frac{1}{3} = 1\frac{1}{3}$ and $6 \times \frac{1}{6} = 1$.)

- Encourage students to show their work concretely using the pattern blocks, pictorially by drawing a picture, and abstractly by writing the multiplication equation that matches each problem.

- Your students' prior understanding of the multiplication of whole numbers includes the commutative property. Once they've had plenty of practice with the problems above, ask, "How can you show me $\frac{1}{2} \times 4$?" ($\frac{1}{2}$ of 4) Continue by asking, "How is this related to $4 \times \frac{1}{2}$?" (4 groups of $\frac{1}{2}$) Give kids time to let this sink in.

- After the eyes start to widen and light up, prod further. Say, "Show me 4 yellow hexagons. What would $\frac{1}{2}$ of the 4 hexagons be?" (2) Next ask, "Now show me 4 groups of $\frac{1}{2}$. How many whole hexagons would that be?" (2) Continue with similar problems and invite your class to contribute to the discussion.

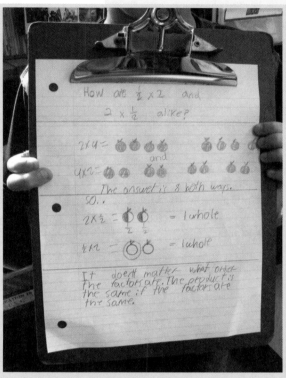

One half of 2 is 1, and 2 groups of $\frac{1}{2}$ equal 1.

While this may seem repetitive to you, it isn't to the children. They need repeated practice with a concept in order to understand it. Thank you, Mr. Piaget.

Write About It: Ask students to illustrate how $\frac{1}{2} \times 2$ and $2 \times \frac{1}{2}$ are related and to explain why this happens.

Extension: Have students use pattern blocks to explore multiplying a fraction by a fraction. Pose the problem, "How can you show $\frac{1}{3} \times \frac{1}{2}$, or $\frac{1}{3}$ of $\frac{1}{2}$?" (A red trapezoid equals $\frac{1}{2}$, and it takes 3 green triangles make 1 trapezoid. Since 1 triangle is $\frac{1}{3}$ of the half, it's $\frac{1}{6}$ of the whole.)

 QUICK TIP

Some children may struggle to remember the value of each pattern block shape. Posting an "anchor chart" in the classroom for all to see, or giving individual students smaller anchor charts, written on index cards, can be very helpful.

5.NF.B.4b Find the area of a rectangle with fractional side lengths by tiling it with unit squares of the appropriate unit fraction side lengths, and show that the area is the same as would be found by multiplying the side lengths. Multiply fractional side lengths to find areas of rectangles, and represent fraction products as rectangular areas.

Math Practices
4 Model with Mathematics
6 Attend to Precision
7 Make Use of Structure

Tiles to the Rescue

When students are asked to find the area of a rectangle whose side lengths include fractional numbers, it's often a real showstopper. Using a concrete model can easily prevent that from happening. You'll want to plan on at least 2 days for this lesson.

To prepare, cut 1-inch square tiles out of paper or card stock. Each student will need 12 to 20 squares, a sheet of 1-inch grid paper, colored pencils, and scissors.

- Begin with small steps. Introduce the concept of multiplying a fraction by a whole number first. Say, "Your first task is to build a $2\frac{1}{2}$ x 3-inch rectangle using your 1-inch square tiles."

- Explain by saying, "Once your rectangle is built, you'll count the total number of squares. That number will give you the product of the multiplication sentence $2\frac{1}{2} \times 3$." If you get confused looks, say, "A 3 x 4 rectangle uses 12 squares and the product of 3×4 is 12." Stop and let kids internalize this tidbit.

- Let students know that they'll need to fold and cut some of their squares in half to create this shape. This may be the first time they've been asked to show "half," and it might not be instantaneous for some. Give time for everyone to place the appropriate squares in a line.

- Once the squares are in place, have kids count the squares. It's easiest if they count whole squares first, and then add the halves. (There are 6 whole squares and 3 half squares. The total is $7\frac{1}{2}$.) Review, "A 3 x 4 rectangle uses 12 squares and $3 \times 4 = 12$. How many squares are in the $2\frac{1}{2}$ x 3-rectangle? Yes, $7\frac{1}{2}$, so the product of $2\frac{1}{2} \times 3$ is $7\frac{1}{2}$."

- Provide students with a few more similar problems. Once they understand this concept using the paper squares, move them to the pictorial stage. Ask them to trace the squares and half squares onto grid paper. Require them to label the measurements and write the equations; for example $2\frac{1}{2} \times 3 = 7\frac{1}{2}$. Remember, when students label, they're attending to precision!

That's enough for the first day. Place the squares, half squares, and grid paper somewhere for safekeeping. Your students may need more

The tiles make it easy for students to see that it would take $7\frac{1}{2}$ 1-inch squares to build a $2\frac{1}{2}$ x 3-inch rectangle.

experience multiplying a fraction by a whole number. So don't rush them. When they're ready, continue by showing them how to multiply a fraction by a fraction.

- Say, "Okay, mathematicians, your task is to build a rectangle that's $2\frac{1}{2}$ inches by $3\frac{1}{2}$ inches using your 1-inch square tiles and your half squares." (They'll need to cut a half square in half to create a fourth, but don't tell them that yet!)

- First say, "Let's put out 1 row of $3\frac{1}{2}$. You used 3 whole squares and then a half, correct? Now, add another row of $3\frac{1}{2}$. You now have 2 rows of $3\frac{1}{2}$. Correct? Now how in the world are we going to show that last *half* row?" Pause, maybe a long pause, as they roll eyes back and forth trying to visualize.

- Continue, "Yes, we need to make that last row half of the row above it. So, there are 3 whole squares that need to be cut in half. But what about that last piece that's already a half? Yes, it needs to be cut in half. What is half of a half? One fourth, right on!"

Your students will probably need more experience multiplying fractions by fractions using paper squares before they begin recording via drawings on the grid paper. But once they do "get it," ask students to trace around their squares, half squares, and quarter squares. Direct the conversation as follows.

This model helps students to see that $3\frac{1}{2} \times 2\frac{1}{2} = 8\frac{3}{4}$.

- Continuing with the example of the $2\frac{1}{2}$ x $3\frac{1}{2}$-rectangle, say, "Count the squares on the grid paper once you're finished drawing. Does your drawing match what you did with the tiles?" Allow time for thinking. Say, "Do you notice that there are 6 whole tiles in all? This represents 6 square inches."

- "Let's look at the tiles that have been cut in half. There are 5 one-half pieces. How many whole tiles can we make with those 5 pieces?" (Use 2 of the halves to make the seventh square-inch tile and 2 more halves to make the eighth square-inch tile.)

- "Now we have eight 1-inch square tiles, one $\frac{1}{2}$ tile, and one $\frac{1}{4}$ tile. How can we combine the $\frac{1}{2}$ and the $\frac{1}{4}$?" (If students cut the $\frac{1}{2}$ piece in half to make 2 fourths they'd have $\frac{3}{4}$. This will give them an area of 8 and $\frac{3}{4}$ square inches.) The last step is for your students to write the matching equation. $\left(2\frac{1}{2} \times 3\frac{1}{2} = 8\frac{3}{4}\right)$

Write About It: Ask, "What if I had a $2\frac{1}{2}$ x $2\frac{1}{2}$-inch square? How many square inches would this square have? $\left(6\frac{1}{4}\right)$ How do you know? Explain your ideas using words and/or pictures."

Paper Plates Show the Way

This investigation gives students the opportunity to discover that multiplying a fraction by a fraction results in a product less than either factor. This generalization is a concept, and it's best if they discover it for themselves. Resist the urge to blurt this out. Remember, whoever is working the hardest is learning the most!

Arrange students in small groups and give each group 4 sets of Paper Plate Fractions, see page 88. Begin by taking time to discuss the fractional sizes of the Paper Plate Fractions, and then write this problem on the board:

Sarah says that when you multiply a fraction less than 1 by a whole number, the product will always be less than the whole number. Judd says this can't be right. Multiplication always gives us a bigger number! Who is right and why? Use the equations $\frac{1}{2} \times 3 = ?$ and $\frac{1}{4} \times 4 = ?$ to prove or disprove your point. You may use the paper plate fractions to show your thinking.

➥ Sarah is right.

- Say, "Let's solve the first equation together. $\frac{1}{2} \times 3$. Let's think about this. If I turn the factors around for $\frac{1}{2} \times 3$ it also means 3 sets of $\frac{1}{2}$. So pull out three $\frac{1}{2}$ pieces from your paper plate fraction bag. What do you have? Yes. Three halves are equal to $1\frac{1}{2}$."

C ▸ P ▸ A
Small Groups

5.NF.B.5 Interpret multiplication as scaling (resizing), by:
5.NF.B.5a Comparing the size of a product to the size of 1 factor on the basis of the size of the other factor, without performing the indicated multiplication

Math Practices
3 Construct Arguments & Critique Reasoning
4 Model with Mathematics
7 Make Use of Structure

Paper plate fraction pieces help give your mathematicians a concrete way to visualize problems.

- Take a deep breath, giving kids time to internalize what they just saw. This may be the perfect time to ask one of your students to restate what just happened so that you can assess if this demonstration hit the mark. Now it's time to look at the same problem from another view.

- Continue, "Let's think about this as $\frac{1}{2}$ of 3 instead of 3 sets of $\frac{1}{2}$. First of all, should we get the same answer? Yes, that's right! Thanks to the commutative property, no matter what way we place the factors in a multiplication sentence, the answer does not change. 4×3 is 12 and 3×4 is 12, right?"

- Give your mathematicians a concrete way to visualize this problem. Say, "Pull out 3 whole plates. Now look at those plates. What is $\frac{1}{2}$ of those 3 plates?" Give a little wait time before continuing with, "Yes, it is 1 and $\frac{1}{2}$! Just like the answer we arrived at with $\frac{1}{2} \times 3$."

- Reread the problem you wrote on the board and say, "Now it's your turn to take the equation $\frac{1}{4} \times 4$ and work this problem out using paper plates. Remember, $\frac{1}{4} \times 4$ means $\frac{1}{4}$ of 4. Or 4 sets of $\frac{1}{4}$. Please use your plates to prove your answer. As a team, be prepared to talk about how you got your answer." (1)

Give your kiddos plenty of hands-on experience using the paper plates to solve similar problems. The goal here is for kids to see a problem such as "$3 \times \frac{1}{3}$" and to visualize 3 one thirds, and understand that 3 one thirds is equal to 1 whole. Visualizing is so important! Be sure your kiddos really understand this with the paper plate fraction tool. Once they "get it" they'll be able to visualize the problems and the fractional pieces—promise!

Write About It: Say, "Sarah is looking at the problem $\frac{1}{8} \times 5 = ?$ Her teacher asks if the answer will be 1, less than 1, or greater than 1. How can Sarah know the answer without calculating?" (Eight eighths make 1 whole. Five groups of $\frac{1}{8}$ will be less than 1.)

Reasoning with Fractions

Here's the crux of this standard:

- If a whole number is multiplied by a fraction = to 1, the product will be = to the whole number.
- If a whole number is multiplied by a fraction > 1, the product will be > than the whole number.
- If a whole number is multiplied by a fraction <1, the product will be < than the whole number.

Plan to spend a part of your math class each day, for 3 days, addressing just 1 concept (column on the chart) at a time. Students need colored pencils and their math journals.

Day 1: Have students create a chart with 3 columns, as shown in the photo, and label only the heading from column 1. (Yes, you could create this chart on your computer for the children, but the meaning behind these words is so important that it's beneficial for kids to write them.)

Continue the discussion. Say, "Fractions with the same number in the denominator and numerator always equal 1. Yes, $\frac{4}{4}$, $\frac{73}{73}$, $\frac{165}{165}$, all equal 1!" To demonstrate draw a rectangle with 5 parts. Say, "This rectangle has 5 parts. The entire rectangle is $\frac{5}{5}$, which is the same as 1 whole. What do we know about any number multiplied by 1? Right, when a number is multiplied by 1, the product is always that number."

Continue by saying, "Let's draw and write a multiplication sentence that demonstrates this." Instruct students to independently draw and write at least 5 examples in the first column of their charts. You may want to meet in small groups for kids who aren't ready to work on their own.

5.NF.B.5b Explaining why multiplying a given number by a fraction greater than 1 results in a product greater than the given number (recognizing multiplication by whole numbers greater than 1 as a familiar case); explaining why multiplying a given number by a fraction less than 1 results in a product smaller than the given number; and relating the principle of fraction equivalence $\frac{a}{b} = \frac{(n \times a)}{(n \times b)}$ to the effect of multiplying $\frac{a}{b}$ by 1.

Math Practices

3 Construct Arguments & Critique Reasoning

4 Model with Mathematics

This lesson helps students discover that this concept-laden standard is actually quite manageable!

Day 2: Ask, "What do we know about fractions where the numerator is greater than the denominator? Yes, those fractions equal more than 1. For example, $\frac{6}{5}$ is more than 1, and $\frac{7}{4}$ is more than 1." Be sure your charges really understand this before continuing.

Ask, "What do you think will happen if you multiply a whole number by a fraction that's greater than 1, such as $4 \times \frac{4}{3}$?" Give your students time to think. "Bingo! The answer will be more than the whole number."

Have students fill in the heading for column 2 on their charts (use the photo as your guide). As a group, come up with a few examples of multiplication sentences showing a whole number multiplied by a fraction greater than 1, and then turn kids loose to add in more examples and illustrations.

Day 3: On this day students fill in the last column on their charts. Say, "If you multiply any number by 1, you'll always get that number. If you multiply a whole number by a fraction that's greater than 1, your answer will always be greater than the whole number." Help them make the leap. Ask, "So, what do you think will happen if you multiply a whole number by a fraction that's less than 1?" Give your kids plenty of think time. "Yes, you'll get a number less than the whole number!"

Provide a visual example. Draw a rectangle. Divide it into thirds. Say, "If I had $4 \times \frac{3}{3}$ complete rectangles, I would get a total of 4 rectangles, right?" Now shade in 2 parts of the rectangle. Ask, "What if I multiplied 4 times $\frac{2}{3}$ of the rectangle? Would I get a total of 4? Of course not; I'd get less than 4."

Ask students to fill in the heading for the last column and write some sample problems. Require students to use illustrations and number sentences to demonstrate this last type of fraction problem.

Painting Turtles

This lesson deals with multiplying mixed numbers and fractions, which is a difficult concept for most students. However, when children use tools, like the paper fraction turtle pieces, they build understanding.

Each student needs 1 set of paper fraction circles cut in wholes, halves, quarters, and eighths (as used in Building Fractions with Box Turtles, page 117).

Display the following problem and let your students know that you'll work through it together:

Cathy painted $\frac{3}{4}$ of her turtle shell green. Cindy put stripes on $\frac{2}{6}$ of the painted part. What fraction of the turtle has stripes?
➡ One fourth of the turtle has stripes.

- Instruct students to use the paper fraction pieces to show $\frac{3}{4}$ of a turtle shell. Point out that the $\frac{3}{4}$ represents three fourths of the whole circle or turtle shell.

5.NF.B.6 Solve real world problems involving multiplication of fractions and mixed numbers, e.g., by using visual fraction models or equations to represent the problem.

Math Practices
1 Solve Problems & Persevere
3 Construct Arguments & Critique Reasoning

Working with the paper fraction pieces helps students see that $\frac{2}{6}$ of $\frac{3}{4}$ is $\frac{1}{4}$.

- Remind students that the task says Cindy put stripes only on $\frac{2}{6}$ of the *painted* part, and that is a portion of the whole turtle. Help students understand that they're starting with a fraction or a portion of the turtle, just $\frac{3}{4}$ of the turtle. Not the whole turtle!

- Kids come to us with so many misconceptions about problems like this one. Go really s-l-o-w-l-y! Be absolutely certain that the children understand the problem starts with a fraction of a whole and moves to a fraction of a fraction. Press on only when you're confident that your kiddos are ready.

- Say, "Place the three $\frac{1}{4}$ pieces on your work space. This is $\frac{3}{4}$." Continue, "Cindy is painting just a portion of the $\frac{3}{4}$." Work slowly and generate a conversation about how they can find $\frac{2}{6}$ of those 3 pieces. Suggest they think about dividing those three $\frac{1}{4}$ pieces into 6 even pieces. Don't fret if they don't get it instantly.

- Someone will light up when she realizes, "Six of the $\frac{1}{8}$ pieces are equivalent to the $\frac{3}{4}$ part of the turtle shell. Two of the $\frac{1}{8}$ pieces are $\frac{2}{8}$, or $\frac{1}{4}$ of the whole."

- Hooray! Someone got it! However, this is not a signal to move on. Be sure each sweet face in your classroom understands this concept. Ask your students to help their classmates "see" what's going on. Don't take it personally; you can say something 100 times, yet some students won't "get" it until they've heard it from a classmate!

- Continue the conversation. Point to the $\frac{2}{6}$ of the portion. Ask, "Are there fraction pieces that you can use to show those 2 sixths?" (Yes, two $\frac{1}{8}$ pieces.)

- Say, "Now look at the pieces. How much of the original *whole* turtle are the two $\frac{1}{8}$ pieces covering? Yes the two $\frac{1}{8}$ pieces are equal to $\frac{1}{4}$. So the answer to $\frac{3}{4} \times \frac{2}{6}$ is $\frac{1}{4}$!"

- You'll want to give your students more practice using the manipulatives to solve similar problems. When they're able to visualize what will happen to the paper fraction pieces *before* they even move them, they're ready to use the traditional algorithm to solve multiplication problems.

The Lion's Share

 The Lion's Share: A Tale of Halving Cake and Eating It, Too by Matthew McElligott

In this story, the animals cut and share Lion's birthday cake until there are only crumbs left!

Students need a copy of the Lion's Share Data Sheet on page 221, an 8 x 16 grid, and colored pencils or crayons.

Read this story to your class once for the pure joy of it, and then pass out the 8 x 16 grids.

- Begin by saying, "Your grid has 128 rectangles, just like the inside cover of this book. The grid represents Lion's cake at the start of the story. Let's leaf through the pages of the story and divide and label the grid according to the size of cake each creature took from Lion's cake."

- Say, "First Elephant ate half of the cake. How can you divide your grid in half?" Allow conversation. If necessary, suggest that students hold their papers horizontally and make a vertical line down the middle of it. Ask students to label and color Elephant's half any color they wish.

- Say, "Hippo ate half of what was left of the cake. How can you divide the remaining half piece of cake in half?" Elicit discussion. Ask kids to label and shade in Hippo's section using a different color. Ask, "What fraction of the whole did Hippo eat?" $\left(\frac{1}{4}\right)$

- Say, "Next Hippo passes half of a fourth of a piece of cake to Gorilla. Draw a line on your grid to divide what's left of the cake in half. Color and label Gorilla's piece. What fraction of the original cake is left?" $\left(\frac{1}{8}\right)$

- Continue in this same manner, asking students to color and label their grids until poor Ant is left with only crumbs for herself and Lion. (Here's how the rest of the cake is divided: Tortoise gets $\frac{1}{16}$, Warthog gets $\frac{1}{32}$, Macaw gets $\frac{1}{64}$, Frog gets $\frac{1}{128}$, and Ant and Lion get $\frac{1}{256}$ or crumbs.) Be sure to ask your students to cut that last rectangle in half!

C P A
Whole Group

5.NF.B.7 Apply and extend previous understandings of division to divide unit fractions by whole numbers and whole numbers by unit fractions.
5.NF.B.7a Interpret division of a unit fraction by a non-zero whole number, and compute such quotients. *For example, create a story context for $\left(\frac{1}{3}\right) \div 4$, and use a visual fraction model to show the quotient. Use the relationship between multiplication and division to explain that $\left(\frac{1}{3}\right) \div 4 = \frac{1}{12}$ because $\left(\frac{1}{12}\right) \times 4 = \frac{1}{3}$.*

Math Practices
4 Model with Mathematics
6 Attend to Precision
7 Make Use of Structure
8 Express Regularity in Repeated Reasoning

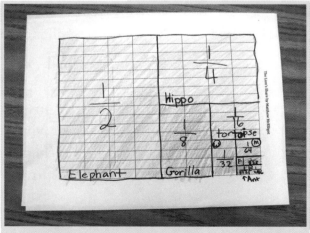
You can easily create this 8 x 16 grid as a chart using any word processing program.

Lion's Share Data Sheet *Use with The Lion's Share, page 133.*

Name: _____ Date: 3-10-14

Record what fraction of the cake each animal ate.

Elephant	Hippo	Gorilla	Tortoise	Warthog	Macaw	Frog	Ant
1/2	1/4	1/8	1/16	1/32	1/64	1/128	1/256
64/128	32/128	16/128	8/128	4/128	2/128	1/128	

Write equations to represent the first 3 ways the cake is cut.

Cut #1: $1 \div 2 = \frac{1}{2}$

Cut #2: $\frac{1}{2} \div 2 = \frac{1}{4}$

Cut #3: $\frac{1}{4} \div 2 = \frac{1}{8}$

Look carefully at each of the equations you recorded in the box above. What generalization can you make about what is happening to the fractions each time the cake is cut? When 128 is the denominator the numerators are half the size of the numerator before. When 1 is the numerator, the denominators are twice the size as the denominator before. (Those are simplest terms)

Which Mathematical Practices did you use to solve this task?

Persevere, look for repeated patterns, precision.

Filling in the data collection sheet helps students easily see the pattern occurring with the numerators and denominators.

- Next pass out the Lion's Share Data Sheet copymaster. The top section asks students to fill in the fraction of cake each animal ate. Students may write fractions with 128 as the denominator. As students work, however, they should recognize that the fractions can be simplified. $\frac{64}{128} = \frac{1}{2}$, $\frac{32}{128} = \frac{1}{4}$, $\frac{8}{128} = \frac{1}{8}$, and so on. Ask them to record those equivalent fractions in the boxes on their Data Sheets as well.

- Next, children are asked to write equations to represent the first 3 ways the cake is cut. At first, the whole cake is cut into 2 pieces. The matching equation is $1 \div 2 = \frac{1}{2}$. Hippo takes the remaining half of cake and divides it into 2 pieces. The matching equation is $\frac{1}{2} \div 2 = \frac{1}{4}$. Now there's one-fourth of the cake left. Gorilla divides that portion into 2 pieces. The matching equation is $\frac{1}{4} \div 2 = \frac{1}{8}$.

- The Data Sheet only requires equations for the first 3 cuts because those numbers are manageable. Of course, some students may want to continue writing equations because the pattern is very apparent to them. Fifth graders must deal with any denominator, so this is excellent practice.

- Finally, students are asked to make a generalization about what they notice happening to the fractions each time the cake is cut (the unsimplified numerator gets smaller by half) and to identify what Mathematical Practices they used to solve this task (mathematical practice 8 sure comes into play here with the repeated reasoning).

Mr. Franklin's Fractions

Most fifth graders are up to their proverbial knickers in US history. The Benjamin Franklin scenario featured in this activity combines math and social studies with imagination and gives your students the chance to work in small groups or independently.

All you'll need is a bit of dramatic flair and a place to post the problem in this activity. Your kids will need paper and pencil, and the fraction tools of their choice. (Pattern blocks and Fraction Rulers, page 102, make excellent choices. If necessary, remind students that tools include drawing a picture or making an area model drawing.)

- Set the stage: "Boys and girls, what if Ben Franklin and 4 of his friends wanted to share 10 apples? That would be easy to figure, right? Each friend would get 2 apples. But what if the number of apples didn't "fit" so easily? What if the 5 men wanted to share 4 apples? Would they each get a whole apple?" (no)

- Your students have been working on fractions for a while now, so give them time to stew about this problem. After some think-time, say, "Let's imagine Mr. Franklin cuts each apple into 5 pieces, 1 piece for each friend. How many pieces would that be? Yes, 20. Twenty *pieces,* not 20 apples. Four apples cut into 5 pieces can be written as $4 \times 5 = 20$."

- Pause and let the apple idea sink in. You might need to draw 4 apples sliced into fifths or sketch simple rectangles cut into fifths. Say, "Each man would get 1 of the fifths from each of the 4 apples, right? This means $4 \times \frac{1}{5} = \frac{4}{5}$ for each man. We knew that each man would have less than 1 whole apple."

- You may want to pose similar problems before turning kids loose on the next scenario.

- Say, "Imagine the year is 1776. Our Founding Fathers are hard at work making plans for the new nation. Benjamin Franklin has requested that a full day of meals be prepared for 7 guests and himself. The servers are to divide the food evenly among the 8 diners. (If they don't eat all their food, Ben will insist they take the food home!) Here's the list of provisions for Mr. Franklin's Feast." Post the following problem:

C P A
*Small Groups,
Individuals*

5.NF.B.7b Interpret division of a whole number by a unit fraction, and compute such quotients. *For example, create a story context for $4 \div \left(\frac{1}{5}\right)$, and use a visual fraction model to show the quotient. Use the relationship between multiplication and division to explain that $4 \div \left(\frac{1}{5}\right) = 20$ because $20 \times \left(\frac{1}{5}\right) = 4$.*

Math Practices
3 Construct Arguments & Critique Reasoning
4 Model with Mathematics
5 Use Tools Strategically

At the feast Ben Franklin and his 7 guests will evenly share: 4 cups of cranberries, 2 cups of green beans, 5 cups of hasty pudding, 11 cups of steamed pumpkin pudding, 9 Indian corn sticks, 3 johnnycakes, 7 pounds of venison, 12 Dolly Madison rolls, 2 Sally Lunn cakes, 1 chicken pot pie, and 68 ounces of wassail. Figure the exact serving size for each man. Please include each division and related multiplication sentence.

➡ Each man was served $\frac{1}{2}$ cup of cranberries, $\frac{1}{4}$ cup of green beans, $\frac{5}{8}$ cups of hasty pudding, $1\frac{3}{8}$ cup of steamed pumpkin pudding, $1\frac{1}{8}$ Indian corn sticks, $\frac{3}{8}$ johnnycakes, $\frac{7}{8}$ pounds of venison, $1\frac{1}{2}$ Dolly Madison rolls, $\frac{1}{8}$ of the total Sally Lunn cakes (or $\frac{1}{4}$ of 1 cake), $\frac{1}{8}$ chicken pot pie, and $8\frac{1}{2}$ ounces of wassail.

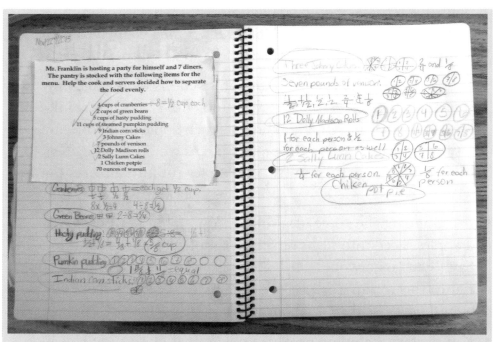

Hearty work calls for hearty fare. The drawings and equations in this student's journal reflect her deep understanding of fractions.

Old-Fashioned Dessert Time

The terms "real world" and "dividing fractions" are in the same sentence in this standard. Is that possible? Of course it is! Have you ever had to cut a recipe in half? What about sharing $2\frac{1}{2}$ pizzas fairly among 5 people? Kids relate to contextual problems much more easily than if we just ask them, "What's half of $\frac{3}{4}$, or what's $2\frac{1}{2} \div 5$." This sweet activity really softens the blow of this tough concept!

The recipes in this lesson come from *The Rumford Complete Cook Book* (pages 148 and 150) published in 1918 (it belonged to a particularly wonderful Nana). All but one of the ingredients is measured in unit fractions, which is what this standard refers to. Your students' task is to figure out how much of each ingredient would be needed to create half of each recipe.

Encourage students to choose a fraction tool (such as the Fraction Rulers on page 102, pattern blocks, or drawing a picture) to help them solve this problem.

You'll want to display just one of these recipes at a time. Go ingredient-by-ingredient, insisting kids have private think-time before calling on anyone to answer and explain their reasoning. If your chefs

C P A
*Whole Group,
Individuals*

5.NF.B.7c Solve real world problems involving division of unit fractions by non-zero whole numbers and division of whole numbers by unit fractions, e.g., by using visual fraction models and equations to represent the problem. *For example, how much chocolate will each person get if 3 people share $\frac{1}{2}$ lb of chocolate equally? How many $\frac{1}{3}$-cup servings are in 2 cups of raisins?*

Math Practices
1 Solve Problems & Persevere
2 Reason Abstractly & Quantitatively
6 Attend to Precision
7 Make Use of Structure

Cutting a recipe in half hasn't changed over time.

struggle when dividing $\frac{2}{3}$ and $\frac{3}{4}$ in half, encourage them to think of $\frac{2}{3}$ as the equivalent fraction $\frac{4}{6}$ and $\frac{3}{4}$ as the equivalent fraction of $\frac{6}{8}$. But… don't do this too quickly; let them tussle.

After working through 1 recipe together, students may have the confidence to divide the other recipe on their own. If you have 1 student who solved the problem with pattern blocks and another who used the Fraction Rulers, ask those children to explain their methods. This will prove that different tools arrive at the same answer. Bon appétit!

<u>Moon Cake</u>
$\frac{1}{2}$ cup butter
$\frac{2}{3}$ cup sugar
2 eggs
$\frac{1}{3}$ cup blanched and chopped almonds
$\frac{1}{3}$ tsp. salt
$1\frac{1}{2}$ cup flour
2 level tsp. baking powder

<u>Ten Minute Cake</u>
$1\frac{1}{2}$ cups flour
$\frac{3}{4}$ cup sugar
2 level tsp. of baking powder
$\frac{1}{4}$ tsp. salt
$\frac{1}{4}$ cup butter
$\frac{1}{2}$ tsp. flavoring extract
2 eggs
1 cup milk

Extension: Ask students to bring in copies of recipes from home. Set up a center with paper, pencils, and a variety of manipulatives. Ask students to select a recipe to cut in half or thirds and copy it onto the paper.

QUICK TIP

If some of your students are struggling, you may want to refer to Full House, page 110 to provide them with extra support.

Measurement and Data

This domain is chock-full of important everyday math skills. Take a moment to consider this fact: No matter what career paths your students may eventually take, you can be certain that they'll need to employ the skills of measurement and data. Food, fuel, paint, tile, mileage, carpet, medicine, altitude, minerals, marathons, and so much more are all measured in specific units.

The concepts presented in this chapter shouldn't be taught from worksheets. They measure up to real-world application. When students are learning about liquid measurements, let them roll up their sleeves and pour. To help them understand area, let kids cover large areas of the classroom with equal-sized squares and then have them count the squares. To help them deeply understand acute and obtuse angles, give kids the time and space to construct a variety of different ramps and then race cars down them. We think engaging, hands-on activities like these are essential for building understanding!

Supporting students as they become proficient and educated consumers of data is critical, too. We're all bombarded daily with information that's presented in the form of charts, graphs, and line plots. The activities you'll find here are meant to help students become adept at reading, organizing, and understanding such data in order to make well-informed, accurate, and meaningful comparisons.

The skills learned in this domain are essential to future activities such as cooking, building, and understanding financial charts and graphs. Pay attention kids!

It's fun and easy to bring the real world into your measurement and data lessons!

GRADE ③

Cluster 3.MD.A Solve problems involving measurement and estimation of intervals of time, liquid volumes, and masses of objects.

C ▶ P ▶ A
Whole Group

3.MD.A.1 Tell and write time to the nearest minute and measure time intervals in minutes. Solve word problems involving addition and subtraction of time intervals in minutes, e.g., by representing the problem on a number line diagram.

Math Practices
1 Solve Problems & Persevere
2 Reason Abstractly & Quantitatively
4 Model with Mathematics
6 Attend to Precision

Get Up and Go!

 Get Up and Go! by Stuart J. Murphy

This story, told from the dog's point of view, allows your students to explore the passage of time using clocks, timelines, and the more abstract written time. Mental math strategies are used to keep a running tab on the total number of minutes that have passed.

Each student needs a clock with plastic hands (handmade paper-plate clocks work just fine too), a copy of the Time Is Passing Data Sheet on page 222, and colored pencils. A document camera, so students can share their work with the class, is optional.

- Pass out the clocks and begin with a review of telling time to the hour, half hour, and minute. You might say something such as,

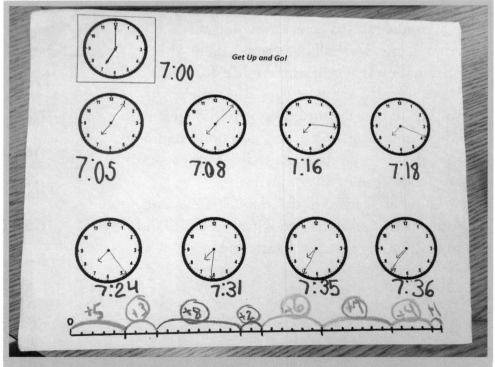

You can see the compatible numbers marked in different colored pencils on the student data sheet pictured here: $3 + 7$, $8 + 2$, and $6 + 4$ make 3 sets of 10. Just add $5 + 1$ to get 36 total minutes.

"The hour hand is short and slow. It takes a whole hour to move from one number to the next on a clock. The minute hand is long and fast like a long-legged runner. It races around the clock, taking only 5 minutes to go from number to number."

- Point out the individual ticks that represent the minutes between each number. You may want to review counting by 5s around the clock, or even have the class sit still and be quiet for 1 minute to set the stage.

- Pass out the Time Is Passing Data Sheet to your students. The first clock on the sheet is already set to 7:00. Instruct students to set their clocks to show 7:00 a.m.

Get Up and Go! is the perfect book for helping students understand elapsed time.

- As you read the story, students use the recording sheet to mark the elapsed minutes on the clocks and timelines. Require students to set the hands on their clocks to show each interval of time too.

- The story begins with 5 minutes to snuggle, followed by 3 minutes to wash, 8 to eat, and 2 for a treat. As the story progresses, ask students to circle the numbers on their timelines that make a "friendly 10" (for example 8 + 2 or 7 + 3) in the same color so they can easily add the minutes as seen in the photo on page 140.

- Pause during the story and invite students to share their work using the document camera. Ask questions such as, "Can you find 2 numbers that make 10?" and "How many minutes have elapsed at this point?" Discuss how to find the total amount of time that's passed using compatible numbers (numbers that are easy to manipulate mentally).

Extension: Stock a center with clocks and adding machine tape and have students create their own timelines. Give the children school situations that require them to figure time to the minute, as well as the time that has passed.

3.MD.A.2 Measure and estimate liquid volumes and masses of objects using standard units of grams (g), kilograms (kg), and liters (l). Add, subtract, multiply, or divide to solve one-step word problems involving masses or volumes that are given in the same units, e.g., by using drawings (such as a beaker with a measurement scale) to represent the problem.

Math Practices
4 Model with Mathematics
7 Make Use of Structure

Build a Liter

Can you imagine using inches to build a quart or a gallon? The metric system gives us something we can't do with customary measurement. In this lesson, your third graders transfer their knowledge of linear measurement (centimeters learned in second grade) to volume as they build a liter. A liter is 1,000 cubic centimeters or 1 cubic decimeter.

Each student needs a sheet of construction paper, a ruler marked in centimeters, a pair of scissors, a pencil, and tape. This fun and easy project will take two days to complete.

The first step is to ask students to draw six 10 x 10-centimeter squares. If your students aren't experienced measurers, this can be a challenge. The photo here shows how students can draw 2 rows of three 10 x 10-centimeter squares. This is usually enough for the first day.

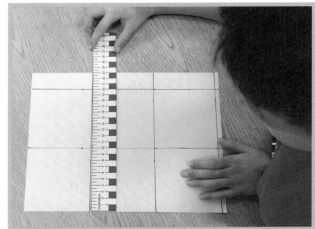

Students can cut six 10 x 10-centimeter squares from a sheet of construction paper. Each 10 x 10-centimeter square is a square decimeter. Link this to area and you have 6 square decimeters. Take the opportunity to discuss these connections!

Put student work away in a safe place.

On the next day, have students cut out and tape their 6 squares to make cubes. Demonstrate these simple steps for making a cube. First tape 4 of the squares in a vertical row. Next tape the remaining 2 squares—1 on the right of the second square and 1 on the left side. Finally, fold the 6 squares into a cube, taping the last 2 faces in place. Ta-da!

Explain to your students that this cube could hold 1 liter of liquid if it were made out of plastic. Stack 2 of these paper cubes next to a 2-liter soda bottle, and compare the sizes. Compare one to a plastic liter container and point out what a liter looks like. Refer to this lesson throughout the year so that these benchmarks become common knowledge.

Tell students to take their time and match the sides exactly when forming their cubes!

C P A

*Whole Group,
Individuals*

3.MD.A.2

Math Practices
4 Model with Mathematics
6 Attend to Precision

Liter Benchmarks

To lend some real-world flavor to this concept, ask students to bring in empty and clean liter-sized containers. Clearly label the size of each container. Seeing the labeled containers will help kids form a mental image of what a 1-liter or 2-liter bottle looks like.

Next, create a center stocked with a variety of containers, including paper cups and small plastic jars and bowls. Post a question such as, "How many of these paper cups would you need to fill to equal 1 liter?"

Have children visit the center and post their guesses using sticky notes. At the end of the day bring 1 liter of water to the whole group. Ask a volunteer to pour the water into paper cups until the liter is empty. Some students will cheer because they were exactly right or very close.

The next day, change the container but ask the same question. For example, "How many of these jars would you need to fill to equal 1 liter?" Count on your students getting better each day, as they look forward to the afternoon "reveal." This center allows students plenty of rich exploration and if you add a tiny bit of food coloring to the water, it's even more fun. You may want to reward the mathematician with the closest guess by asking him to do the reveal!

The daily reveal gives children benchmarks for understanding the liter measurement.

GRADE **3**

Cluster 3.MD.B Represent and interpret data.

C ▸ **P** ▸ **A**
Whole Group

3.MD.B.3 Draw a scaled picture graph and a scaled bar graph to represent a data set with several categories. Solve one- and two-step "how many more" and "how many less" problems using information presented in scaled bar graphs. *For example, draw a bar graph in which each square in the bar graph might represent 5 pets.*

Math Practices
2 Reason Abstractly & Quantitatively
3 Construct Arguments & Critique Reasoning
5 Use Tools Strategically

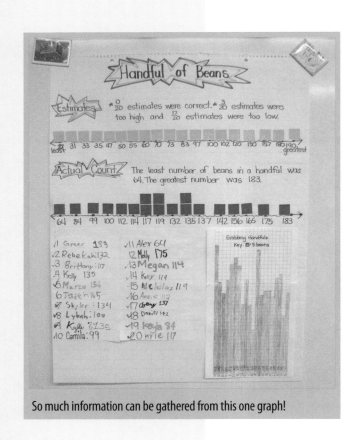

So much information can be gathered from this one graph!

Handfuls of Beans

 Jack and the Beanstalk (at least two different versions)

With a minimum of 20 ounces of any type of dried beans, 4 salad-sized bowls, 1-inch tiles, a half-sheet of graph paper for each person, and the folktales, this lesson connects bar graphs to the CCSS Literacy standard RL.3.2! (That standard includes recounting stories such as folktales and determining the story's central lesson or moral.)

During your literacy time, read at least 2 of the many different versions of the *Jack and the Beanstalk* folktale to your class. Compare and contrast the stories and look for the story's lesson. Now you're ready for the math!

• Let your kids watch as you fill 4 bowls with beans. Ask each child to estimate the number of beans he thinks he can hold in 1 hand. Have students record their estimates in their math journals.

• After all estimates are written, model how to create a class graph of estimates. Draw a line on a piece of chart paper or whiteboard. Since you're creating a line plot, be sure that you have arrows at both ends. You may want your kids to create their own graphs in their math journals, or just watch you.

• Ask students to determine the least estimate. Label that minimum number on one end of your line. Next have students determine the greatest estimate and add that number to the other end of the line.

• Fill in the other numbers between these 2 points on the line, keeping the intervals as equal as possible. Snag this time as an opportunity to talk about intervals and the importance of trying to make the space between the intervals equal.

- Call out the numbers on the line in order. Ask kids to raise their hands when the number you call matches their estimate. Place an *equal-sized-square* (e.g., a 1-inch tile or paper square) above the numbers that represent the estimates in your class. For example, if 4 kids estimated 23, there will be 4 squares above the 23. Tell your children that each square represents 1 person's estimate.

- Now the fun begins! Pass out the bean-filled bowls to different parts of the class-room. Let each student reach into a bowl, one at a time, and grab as many beans as he can hold in 1 hand.

Writing about an activity helps a child internalize what was learned.

- Each child counts the total number of beans in his handful and records that number on a class master list (this is simply a piece of chart paper listing each student's name). Once the beans are counted and returned to the bowl, another classmate repeats this process. (This may take a while, so it should be done while kids are working on another independent task.)

- Once all of the numbers in each handful get recorded on the class master list, pass out graph paper and say, "We'll record everyone's total handful count. Each of you will have a column on the graph. Can anyone think of an efficient way to record the totals, other than filling in 1 graph square for each bean held?"

- Suggest kids fill in 1 graph square to represent 5 beans. Say, "So, if you were able to hold 22 beans, then you'd fill in 4 complete graph squares and part of the fifth square. This is different than how we filled in the class estimate graph, where each square equaled one person's estimate."

- Using initials, give everyone a column indicating his handful. Move slowly as you and the children record names and fill in graph squares. Discuss who has the most and the least, as well as the differences. Ask students, "What other math information can we gather from this graph?"

Write About It: Have children record their estimates and actual counts from the graph, along with what math information they have learned from the graph.

Variation: Use Unifix cubes, marbles, popcorn, or centimeter cubes instead of beans.

C P A

*Whole Group,
Individuals*

3.MD.B.4 Generate measurement data by measuring lengths using rulers marked with halves and fourths of an inch. Show the data by making a line plot, where the horizontal scale is marked off in appropriate units—whole numbers, halves, or quarters.

Math Practices
1 Solve Problems & Persevere
5 Use Tools Strategically

Hand It to Me!

Every child will "hand-deliver" data for this classroom line plot! They'll also be sharpening their skills as they practice measuring and recording to the nearest quarter of an inch.

Each student will need a ruler, pencil, and piece of paper (or math journal). You'll want to use a document camera to demonstrate how to measure to the nearest quarter inch. If you don't have one, then teach this part of the lesson to small groups of students at a time.

- Point out that $\frac{1}{4}$, $\frac{1}{2}$, $\frac{3}{4}$, 1, $1\frac{1}{4}$, $1\frac{1}{2}$, and so on are considered quarters. Use a ruler to measure a pencil and say something such as, "See how this pencil is between $4\frac{1}{2}$ inches and $4\frac{3}{4}$ inches, but it's closer to $4\frac{1}{2}$ inches? So $4\frac{1}{2}$ is its measurement to the nearest quarter inch."

- After demonstrating how to measure at least 4 items to the nearest quarter inch, tell the class that they'll use their rulers to measure their hands to the nearest quarter inch. Explain that they'll measure from their longest finger to the base of their hand, not the wrist.

- To show what you mean by "longest finger" and "base of hand," measure your own hand. If, your hand doesn't measure to an exact quarter of an inch, you'll have another opportunity to demonstrate finding a measurement to the nearest quarter inch!

- Once your children have measured their hands, have them record their measurements on a piece of chart paper. Now it's time for your students to create a line plot of this data.

- The first order of business is to survey your kiddos to find the longest hand; each student's line plot will need to be at least that length.

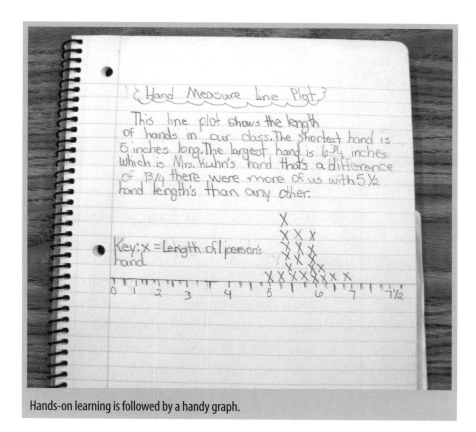

Hands-on learning is followed by a handy graph.

- Instruct children to use their rulers to draw a line that begins with a 0 and ends at least a quarter of an inch past the measurement for the longest hand.

- Give students time to add tick marks to indicate each whole, half, and fourth of an inch. Ask them to label each of those measurements.

- Tell kids to reference the chart paper listing the class data, and place equal-sized "Xs'" over the measurement on the line plot for each classmate's hand length.

Write About It: Ask each student to write at least 3 sentences describing what her line plot shows.

Variation: Have children measure their feet or their shoes and record that information on a line plot.

Cluster 3.MD.C Geometric measurement: understand concepts of area and relate area to multiplication and to addition.

Whole Group

3.MD.C.5 Recognize area as an attribute of plane figures and understand concepts of area measurement.

3.MD.C.5a A square with side length 1 unit, called "a unit square," is said to have "one square unit" of area, and can be used to measure area.

3.MD.C.5b A plane figure which can be covered without gaps or overlaps by *n* unit squares is said to have an area of *n* square units.

Math Practices
2 Reason Abstractly & Quantitatively
3 Construct Arguments & Critique Reasoning
4 Model with Mathematics
6 Attend to Precision
7 Make Use of Structure

Bigger, Better, Best

Bigger, Better, Best! by Stuart J. Murphy

This is the perfect book to help students understand area. The story begins with a typical brother and sister argument. Both siblings think that they have the biggest and best of everything. When their parents decide to buy a new house, this gives them new rooms and windows to compare and argue about. Now who has the biggest and the best?

Gather your students at tables with plenty of space to work. They'll need 1-inch square tiles and 1-inch grid paper. A document camera is optional, but it makes it easy for everyone to see the pictures as you read the story and model the area that corresponds to the windows in the book.

- First read the book to your students for pure enjoyment. The second time through, stop at page 14. This is where the mother tries to help settle her children's argument by giving them square sheets of paper to cover their windows.

- Tell your students that they'll mimic the characters in the book using their square tiles and grid paper. Remind students that as they place their square tiles on the grid paper, they may not have any gaps or overlaps.

- Just as Jeff and Jenny tape the squares on the windows, have your students model 3 rows of 4 concretely with their tiles on the graph paper.

- Ask, "How many square units did it take to cover Jeff's window?" (12) Say, "Can you give me an addition sentence to match your picture?" (4 + 4 + 4 or 3 + 3 + 3 + 3) Continue by saying, "You have 3 rows of 4. Can anyone tell us a multiplication equation that matches this model?" (3×4)

- The next step involves each student drawing and labeling a concrete model of Jeff's window. The 1-inch grid paper will help students attend to precision. Have students label their diagrams as shown in the photo.

- Next the children study Jenny's window. Show students the picture on page 16. The illustration shows the beginning of 2 rows. Ask, "How long would each row have to be for Jenny's and Jeff's windows to have *equal* areas? Turn and talk to your neighbor." Give students time to think about this and build it with their tiles. (It would have to be 2 rows of 6.)

- Challenge students by asking, "How long would each of Jenny's rows have to be for her window to have a *smaller* area than Jeff's window?" (2 x 5 or less) And, "How long would each of Jenny's rows have to be for her window to have just 2 *more* square units than Jeff?" (2 x 7) Continue to allow time for explorations and discussion with each question.

- Show students page 17 and let them see what the storybook children found out. (Jenny has 2 rows of 6—exactly the same as her brother!) Have students draw this window on their graph paper and label it as they did the first window.

- You may want to stop here and bring this book out later to find out who has the larger bedroom. The possibilities are endless!

Write About It: Challenge students to create more than 1 window with an area of 18 square units. Have them draw and label their solutions. (Possible solutions: 3 x 6, 2 x 9, 1 x 18)

Variation: Stock a center with the tiles and graph paper. Post a question such as, "How many different rectangular windows can you make with an area of 16 square units?" (Possible solutions: 4 x 4, 2 x 8, 1 x 16)

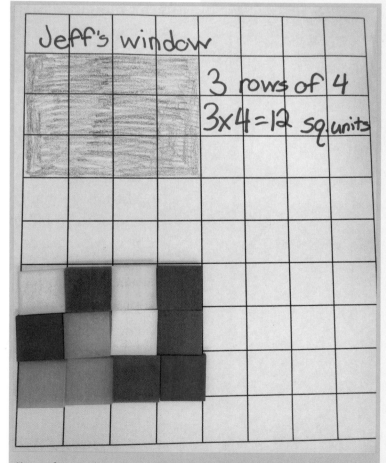

Jeff's window

3 rows of 4
3 x 4 = 12 sq. units

You can formatively assess your students as they model and draw each window.

Small Groups

3.MD.C.6 Measure areas by counting unit squares (square cm, square m, square in, square ft, and improvised units).

Math Practices
1 Solve Problems & Persevere
5 Use Tools Strategically
7 Make Use of Structure

Cover It!

 Bigger, Better, Best! By Stuart J. Murphy

Did you notice that this standard deliberately does *not* require using a formula to determine area? Instead, it focuses on understanding the importance of using a same-sized square to repeatedly cover an area. Students learn to measure area by simply counting the units covering that area.

To begin to "cover" this standard, set up 6 or more stations where small groups of students measure different areas using a variety of improvised measuring units. Improvised units can be equal-sized square tissue boxes, newspapers cut into squares, square-bottomed milk cartons, or square-shaped crackers or cereal pieces.

Tell students that they're going to cover areas just like the characters in the story *Bigger, Better, Best!* did. The storybook characters used large pieces of paper; students are going to use other units. Below are suggestions for several different areas to measure, as well as units to use for measuring (improvised and not), to help develop this understanding:

- ⊙ If the floor in your classroom is made of 1-foot tiles, mark off an area of the floor with tape. Instruct students to count the 1-foot tiles within that area.
- ⊙ Set out one of your largest picture books. Ask students to count the number of 1-centimeter cubes it takes to completely cover the book.
- ⊙ Use tape to mark off a rectangle on the floor that goes from one end of the room to the other end. Provide students with square newspaper pages. (Cut each newspaper sheet into the largest possible square.) Challenge students to count the number of newspaper squares it takes to cover the rectangle.
- ⊙ Designate a student's desk as a special center. Ask children to count the number of 1-inch tiles needed to cover the desk.
- ⊙ Clear off a rectangular table. Require students to count the number of milk cartons (or any other "improvised" unit) it will take to cover it.
- ⊙ Cordon off an area by the sink. Ask students to count the number of 4-inch paper squares needed to cover the area.

Using same-sized units to cover an area shows what the term "square" means in measuring area.

⊙ Tape off an area on your desk (students think it's so fun/adventurous to work at *your* desk!) and have them count the number of graham crackers it takes to cover that area.

Don't be surprised if some of your students discover that they can use their newly learned multiplication skills to arrive at a more efficient answer, but this is a concept-building activity! Let the kids discover and explain this concept to you.

Write About It: Ask students to make a generalization about the number of units they counted and the size of the unit. (You want students to recognize that smaller units require more individual units to cover an area and larger units require fewer.)

Variation: If you're game and you know that your students can manage their excitement outside of the classroom, take the class or invite small groups to the hallway to measure a much larger area by counting 1-foot tiles or using many more newspaper squares.

3.MD.C.7 Relate area to the operations of multiplication and addition.
3.MD.C.7a Find the area of a rectangle with whole-number side lengths by tiling it, and show that the area is the same as would be found by multiplying the side lengths.

Math Practices
1 Solve Problems & Persevere
7 Make Use of Structure

There's a Formula for That!

We want our students to be mathematically efficient. However, we want that efficiency to be rooted in meaning and not just made up of "short-cuts" learned by rote. After all of the counting of repeated units in the preceding activity, it's time to discuss a more efficient method for determining area.

Students will need tiles, 1-centimeter grid paper, and colored pencils.

- Introduce this new strategy with your kids close to you on the floor or watching you demonstrate from the document camera. Say, "Boys and girls, mathematicians have developed an easier way to know the number of tiles covering an area than by counting all the tiles. After all, if I wanted to know how many tiles are in the hallway or office or on the cafeteria floor, it could take a long time to count."

- Build a 5 x 3 rectangle with tiles and say, "See? There are 5 rows of tiles. And there are 3 tiles in each row. If I count them all, 1, 2, 3… I get 15 tiles. But another way to figure this out is to think '5 sets of 3' or '5 times 3.' The answer to 5×3 is 15, so there are 15 tiles in all!"

The tiles make it easy for students to visualize the area of a rectangle. More importantly, it also helps them to understand why length multiplied by width determines area.

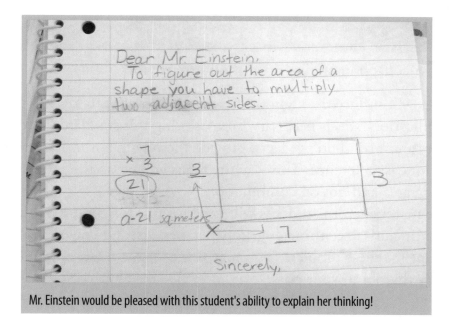

Mr. Einstein would be pleased with this student's ability to explain her thinking!

- Keep demonstrating. Make more rectangles composed of tiles. Calculate the rectangle's area by multiplying and then proving the answer by counting the tiles.

- Divide the class into small groups. Ask students to use tiles to determine the number of tiles needed for rectangles such as 6 x 2, 5 x 5, 4 x 5, and 3 x 6. Insist they place the tiles needed before giving you the answer. Don't rush this process! We're in this for long-term understanding, not short-term "right answers."

You may choose to save this final step for the next day's lesson.

- Pass out 1-centimeter grid paper and ask each student to color rectangles composed of several grid squares. Tell your scholars to include the multiplication sentence alongside each colored rectangle they create to show its area.

- Differentiate the instruction. If you have students who are struggling, you may want to provide them with the rectangle specifics such as 5 x 6 or 5 rows of 6. Strong, independent workers will enjoy knowing that they can go ahead and fill their page with rectangles.

Write About It: Instruct students to write a note in their math journals to Mr. Einstein that explains how to find the area of a rectangle in an efficient manner. Require students to give some examples and label their illustrations.

3.MD.C.7b Multiply side
lengths to find areas of
rectangles with whole-number
side lengths in the context of
solving real world and mathematical problems, and represent
whole-number products as
rectangular areas in mathematical reasoning.

Math Practices
1 Solve Problems & Persevere
6 Attend to Precision
7 Make Use of Structure

Famous People & Places

 A collection of fairy tales or nursery rhymes

Giggles will surface throughout this lesson. Needless to say, knowing multiplication facts is necessary for mastering this standard. If your charges aren't solid on the facts, provide them with a multiplication chart for easy reference. This way the focus will be on the process of figuring area and not on computation.

The materials your students will need for this tongue-in-cheek activity include 1-centimeter grid paper, pencils, and crayons. To set the stage for these problems you may want to read a few fairy tales or nursery rhymes to your class.

- Begin by saying, "Girls and boys, I have a list here of famous people, and I've managed to find the measurements of their homes or work areas. How cool is that! Your task is to take this information and determine the area for each space."

- Continue by saying, "Can anyone refresh us on the way to find the area of a space?" Give time for students to articulate their thinking. Say, "Yes, we can cover the area with square inches, square feet, square yards, square centimeters, or square meters. We can also multiply the number of units on each side to determine the area."

- Pass out the 1-centimeter graph paper and say, "I want you to draw the area I'm describing on the graph paper. Each centimeter square on the paper will be equal to one square meter."

- You may want to read the following problems character-by-character, or you might prefer to display them on chart paper

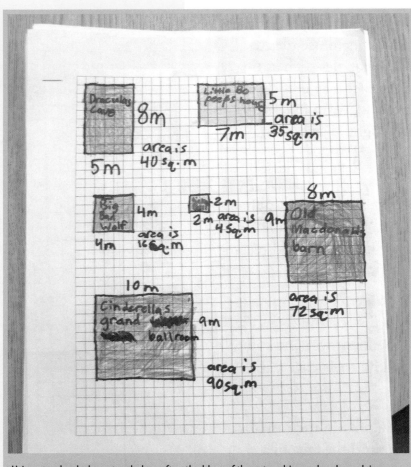

Using storybook characters helps soften the blow of these toughies and makes solving area fun!

or the board. For each problem, require students to draw the area, label the measurements, and write the matching math sentence.

- After you've instilled giggles and inspired your mathematicians, invite them to think of a famous character and a possible chamber, hut, or room for that character. Say, "Write a simple sentence about the character's room similar to the ones I read to you, and then draw the room, label the measurements, write the math sentence, and figure the area. Be clever and have fun! I can't wait to read about your characters!"

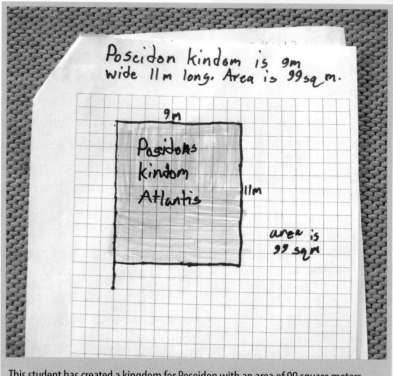

This student has created a kingdom for Poseidon with an area of 99 square meters.

Dracula's cave is 5 meters by 8 meters. What is the area of that cave? Draw a rectangle that shows the floor of Dracula's cave.

➥ 5 meters × 8 meters = 40 meters

Little Bo Peep's cottage is 7 meters by 5 meters. What is the area of her cottage? Draw a rectangle that shows the floor of Bo Peep's cottage.

➥ 7 meters × 5 meters = 35 meters

Big Bad Wolf's hut is 4 meters by 4 meters. What is the area of his home? Draw a rectangle that shows the floor of the Wolf's home.

➥ 4 meters × 4 meters = 16 meters

Old MacDonald's barn is 8 meters by 9 meters. What is the area of Mr. MacDonald's barn? Draw a rectangle that shows the floor of his barn.

➥ 8 meters × 9 meters = 72 meters

Cinderella's grand ballroom at the castle is 10 meters by 9 meters. What is the area of her grand ballroom? Draw a rectangle that shows the floor of this beautiful room.

➥ 10 meters × 9 meters = 90 meters

C ▶ P ▶ A ▶
Whole Group, Pairs

3.MD.C.7c Use tiling to show in a concrete case that the area of a rectangle with whole-number side lengths a and $b + c$ is the sum of $a \times b$ and $a \times c$. Use area models to represent the distributive property in mathematical reasoning.

Math Practices
1 Solve Problems & Persevere
3 Construct Arguments & Critique Reasoning
4 Model with Mathematics
6 Attend to Precision

Distribute and Conquer

This standard asks students to understand that a rectangle with the dimensions of 5 x 7 inches, for example, is equivalent to 2 rectangles that measure 5 x 5 inches and 5 x 2 inches. To understand this, students need to be able to decompose the 7 into 5 and 2.

You'll start this lesson with a whole group instruction and task, move to partner exploration, and then conclude with a class discussion. Students will need tiles, 1-inch grid paper, and crayons. To begin, post this problem:

Laurie and Pat are having a viable argument. Pat has used 1-inch tiles to build a rectangle that's 5 x 7 inches. Laurie has built 2 rectangles with her 1-inch tiles. One is 5 x 5 inches and the other is 5 x 2 inches. Pat says that Laurie's 2 rectangles have a combined area that's equal to her 1 rectangle. Laurie doesn't agree. Who is correct and why?
➡ Pat is correct because both rectangles have an area of 35 square inches.

Ask children to work with a partner. Using the tiles, instruct one partner to build Pat's rectangle and the other partner to build Laurie's 2 rectangles. Give your mathematicians plenty of time to build and then discuss their findings with one another before turning this into a large class discussion.

If conversations lag, use these questions to generate more math talk:

⊙ What did you find when you compared the area of the 5 x 7-inch rectangle to the combined area of the other 2 rectangles? (They were the same.)
⊙ Why does this happen? (Answers should indicate an understanding of the distributive property.)
⊙ Could you take Pat's rectangle and make it look like Laurie's 2 rectangles? (Children can take the 5 x 7-inch rectangle and pull the tiles apart to create a 5 x 5-inch rectangle and a 5 x 2-inch rectangle as shown in the photo.)

Pass out the 1-inch grid paper. Have students each outline a 5 x 7-inch rectangle. Next instruct each student to decompose the

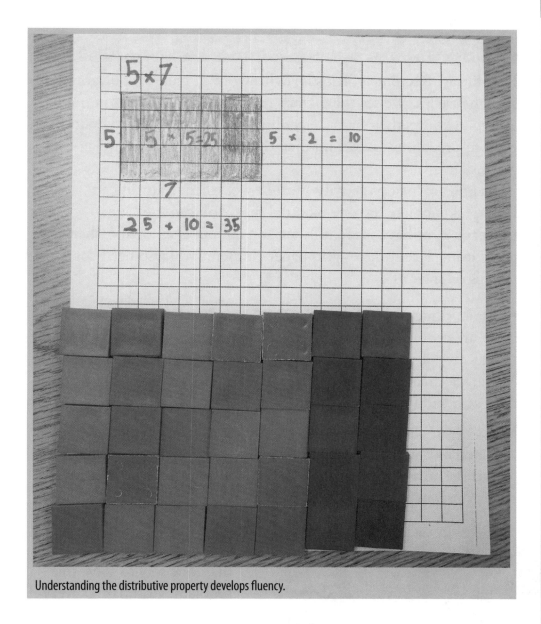

Understanding the distributive property develops fluency.

rectangle by coloring a 5 x 5-inch section of it red and a 5 x 2-inch section of it blue. Ask students to label the red part "5 × 5 = 25" and to label the blue part "5 × 2 = 10." Have them use a crayon to connect the 25 (area) + 10 (area) = 35. This helps the children see that area is additive.

C P A

*Whole Group,
Small Groups*

3.MD.C.7d Recognize area as additive. Find areas of rectilinear figures by decomposing them into non-overlapping rectangles and adding the areas of the non-overlapping parts, applying this technique to solve real world problems.

Math Practices
1 Solve Problems & Persevere
4 Model with Mathematics
6 Attend to Precision
7 Make Use of Structure

A Rectilinear Place to Play

Rectilinear? Rectilinear! Kids love this word! It names a polygon with all right angles. We see these kinds of shapes everywhere, but not everyone knows their name, "rectilinear." In this lesson, students first explore a rectilinear pool and then they create a rectilinear playground.

To prepare for this activity, create a worksheet featuring an L-shaped pool like the one pictured here. You can use Dynamic Paper (see page 9) or have students draw and cut out their own L-shaped pools using 1-centimeter grid paper. Students will also need colored pencils or crayons.

- When students have their rectilinear pools, tell them that each square centimeter represents 1 square meter. Point out to students that they already know how to find the area of a rectangle, so to find the area of this pool they just need to partition it into rectangles and add their areas together.

- Give your charges time to divide the pool into rectangles. Have them color each rectangle using a different color. Ask students to find the area of each colored section and to record their equations and answers as shown in the photo.

Have students decompose the rectilinear shape into rectangles first. Remind kiddos that when they add the area of each rectangle, they'll find the total area of the pool. It's *that* easy!

- The cool thing about this activity is that there's more than one way to solve this problem. Comparing everyone's methods can be fun! If your students have different ways to measure the same shape, it means your children feel free to work problems in a variety of ways. Celebrate!

This next problem is a bit more complex. Each student must create his own unique rectilinear shape, and then he must figure its area.

- Say, "The principal is planning to build a playground. She has asked the third graders to design a rectilinear playground using only 1 sheet of 1-centimeter grid paper." Pass out the 1-centimeter grid paper and tell students that each square centimeter represents 1 square meter.

- Continue by saying, "The padded area of the playground must be at least 300 square meters but no larger than 400 square meters. She needs your help to design this area! Once you've cut out your rectilinear playground shape, color each rectangle and calculate the area."

- Final designs can launch great mathematical conversations. They can also be used as formative assessments. Simply have classmates swap their shapes and solve each other's rectilinear challenges.

GRADE **3**

Cluster 3.MD.D Geometric measurement: recognize perimeter as an attribute of plane figures and distinguish between linear and area measures.

Measurement Detectives

Everyone loves a good mystery! In this lesson, your Measurement Detectives (aka your students) are shown a rectangle and given 2 clues. Clue #1 is either the rectangle's perimeter or its area. Clue #2 is the length of 1 side of the rectangle. With just that information to go on, the detectives must determine the lengths of the other 3 sides of the rectangle.

In order for your students to be able to solve these problems, they must know that parallel sides of a rectangle have equal lengths. They also need to understand the equations for figuring perimeter and area. The thinking is complex and it relies on a solid understanding of perimeter and area.

To get started, all you need is a whiteboard or document camera.

- Begin by saying, "Boys and girls, I love a mystery, and we have several mysteries to solve today. You're going to be Measurement Detectives! I'll show you a rectangle and give you 2 clues about the rectangle. The first clue will be the perimeter of the shape. The

Individuals

3.MD.D.8 Solve real world and mathematical problems involving perimeters of polygons, including finding the perimeter given the side lengths, finding an unknown side length, and exhibiting rectangles with the same perimeter and different areas or with the same area and different perimeters.

Math Practices
1 Solve Problems & Persevere
6 Attend to Precision
7 Make Use of Structure

second clue will be the length of just 1 side. You'll need to determine the lengths of the other 3 sides."

- Review perimeter by drawing a rectangle. Label it 8 meters long and 3 meters wide. Write, "Perimeter = 22 meters."

- Say, "This rectangle has a perimeter of 22 meters, and its measurement is 8 x 3 meters. One side is 8 meters and the side parallel to it is also 8 meters. The other side is 3 meters and the side parallel to it is also 3 meters. When I add 3 + 3 + 8 + 8, the total is 22. So the perimeter is 22 meters. Correct? Would someone like to rephrase what I just said?" If you get some blank stares, go back and review with another rectangle.

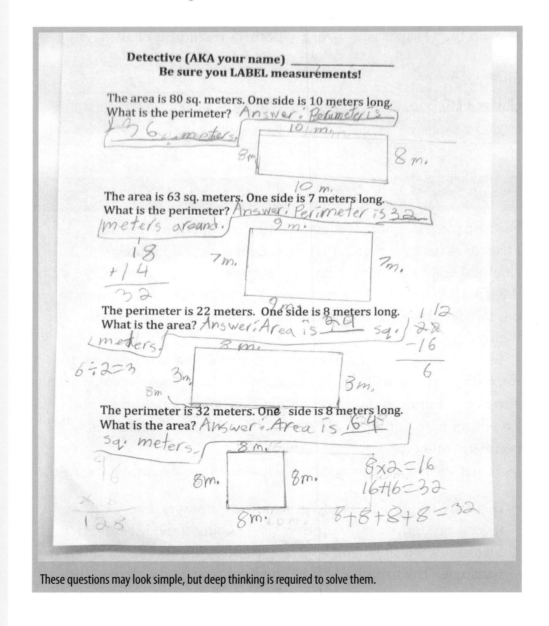

These questions may look simple, but deep thinking is required to solve them.

- Say, "So, if my clues were that the perimeter was 22 and one side was 8, I'd know that the other side must be 8. Next I'd think 8 + 8 = 16 and 22 − 16 = 6. Half of 6 is 3, so 2 sides of the rectangle are 3 meters and the other 2 sides are each 8 meters."

- Go 1 rectangle at a time. Give students the measurement of 1 side of a rectangle and its perimeter. Give your detectives private think-time to solve each mystery. Expect children to explain their thinking. Invite them to answer questions or comments raised by their classmates.

- After some successes (and this may well be a few days later), crank up the lesson by changing Clue #1. Instead of giving the perimeter, give the area of the rectangle. Clue #2 remains the length of one side.

- Discuss how area is the product of length × width. Say, "If we know the area is 32 square units and 1 side is 4 inches, then we can figure out that the other side must be 8 inches. Now what's the perimeter? Yes, 8 + 8 + 4 + 4 or 24 inches!"

Figure the perimeter and/or area of several rectangles ahead of time and you will be ready with lots of rich math content problems that may be used for several days in a row of short "mystery math doses."

Extension: Have students create an interactive bulletin board filled with measurement mysteries. Each student folds a sheet of paper and draws a rectangle on the cover, listing the length of 1 side and the perimeter (or area). When the paper is lifted, the same rectangle is drawn with all measurements and the area (or perimeter) labeled. That way other detectives who solve the problem can check the answer.

Whole Group,
Individuals

3.MD.D.8

Math Practices
1 Solve Problems & Persevere
3 Construct Arguments &
   Critique Reasoning
7 Make Use of Structure

Does Perimeter Determine Area?

This investigation starts with a thought-provoking problem for the class to ponder. Next students move on to independent practice to deepen their conceptual understanding of how perimeter and area are related.

Each child will need 1-inch grid paper, a 16-inch-long piece of string, a ruler, a pair of scissors, one small piece of masking tape, and colored pencils.

- Pose these problems to your young Einsteins and let them think for a while: "Does the perimeter of a shape have anything to do with the area?" and "Do all shapes with equal perimeters have an equal area?" Let them mull these questions over. Don't take answers too early and don't let them know if they're correct or not. There will be time for this at the end of the lesson.

- Pass out the strings. Ask students to find something they can measure by wrapping their piece of string around it. For example, a child might measure a book, a pencil box, his hand, or a folder. Once a student has found an item, ask him to place his string around the item's perimeter and then cut any extra string so that the ends meet *exactly*.

- Demonstrate with a pencil box. Place the string around the box and cut it exactly where the ends touch with no overlapping. Say, "See how this string fits exactly around the box. The string is now equal to the perimeter of the box."

- After a student cuts his string, have him measure the string and record its length. Next demonstrate how to use a small amount of masking tape to secure the 2 string ends together, placed *exactly* end-to-end. Explain, "Now you each have a loop that's equal to the perimeter of the shape you measured."

- Say, "Place your string on your grid paper. Can you form your string into a square? A rectangle? A different-sized rectangle? With each shape you create, what's happening to your perimeter? What about the area?" (Your mathematicians should notice that while the perimeter is constant, the area is changing with each new shape.)

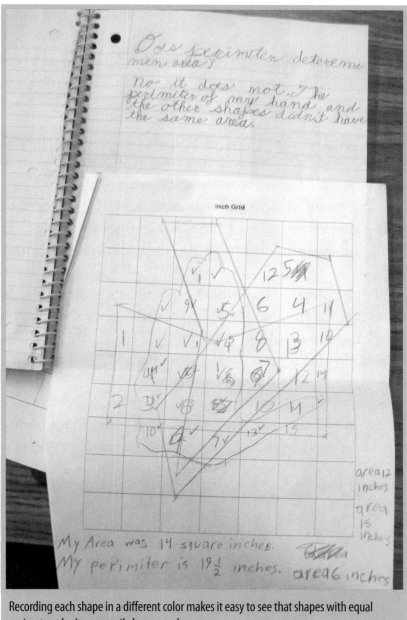

Recording each shape in a different color makes it easy to see that shapes with equal perimeters don't necessarily have equal areas.

- The grid paper is ideal for counting area. Ask students to draw, as accurately as possible, each new shape they create with their loop of string.

Write About It: Ask students to explain what they've learned about perimeter and area, giving specific examples. (You hope they'll record that perimeter doesn't determine area, and shapes may have the same perimeter but not the same area.)

GRADE ④

Cluster 4.MD.A Solve problems involving measurement and conversion of measurements from a larger unit to a smaller unit.

4.MD.A.1 Know relative sizes of measurement units within one system of units including km, m, cm; kg, g; lb, oz.; l, ml; hr, min, sec. Within a single system of measurement, express measurements in a larger unit in terms of a smaller unit. Record measurement equivalents in a two-column table. *For example, know that 1 ft is 12 times as long as 1 in. Express the length of a 4 ft snake as 48 in. Generate a conversion table for feet and inches listing the number pairs (1, 12), (2, 24), (3, 36), …*

Math Practices
4 Model with Mathematics
7 Make Use of Structure

Hamster Golf

As soon as you say, "Today, boys and girls, we'll be playing Hamster Golf," you're guaranteed to have their attention! The object of this entertaining game is to estimate the distance from one hole to the next. Along the way, students must also measure, convert, compute, and compare.

To make the holes, cut out nine 3-centimeter-diameter circles from paper. To create the flags, fold ten 10 x 8-centimeter sheets of paper or cardstock in half. Label the flags "Start," "1," "2," "3," "4," "5," "6," "7," "8," and "9." Each team of 3–4 students will need pencils, a sheet of paper to keep score, and a measuring tape.

To set up the golf course, arrange the paper holes around an area of your classroom or hallway. Remember, this is a course for hamsters, so don't make it too big a space! Tape each hole in place and post a

You may want to find fun images of hamsters on the Internet to jazz up your flags, but that's not necessary; simple numbers are fine.

flag next to it. You'll want to make sure children can see the distances between the numbered holes.

Divide your golfers into teams of 3 or 4. Explain that the object of the game is to estimate the shortest distance from one hole to the next hole. Because the circles are 3 centimeters in diameter, you might want to encourage your students to measure from the closest point on one hole to the closest point on another hole, instead of measuring from the center of each hole. Students can work with customary or metric measurement for this game; it's your choice.

Let's say you choose metric measurement as the focus for the lesson. In turn, each team member estimates the distance from the "Start" flag to the hole labeled "1." After the golfers record their estimates on their team's score sheet, the team works together to measure the actual distance.

HAMSTER GOLF

Hole	Estimate	ACTUAL		
		mm	cm	m
1	56 cm	590	59	0.59
2	110 cm	1550	155	1.55
3	35 cm	510	51	0.51
4	145 cm	1740	174	1.74
5	115 cm	1430	143	1.43
6	60 cm	1000	100	1.00
7	128 cm	1390	139	1.39
8	150 cm	1590	159	1.59
9	50 cm	660	66	0.66

This ridiculously fun game is filled with building benchmarks for unit length.

The actual measurement is recorded in meters (for example, 3.2 meters) and again in centimeters (302 centimeters). Of course, if your students are using customary measurement, then they'll record in feet and inches.

Each team decides which team member had the closest estimate for each hole. This is a mental math challenge. The player with the closest estimate to the correct measurement on each team is the winner of that hole. The players on each team who win the most holes are the winners of the game.

Play this game often throughout the year. As your mathematicians become more fluent with measuring, they'll become better at the game and more competitive with their classmates. After all, who wouldn't want to win at Hamster Golf?

Write About It: Ask kids to share what most surprised them about measuring. (Often students are surprised at how small a centimeter is and how long a meter is. Even though they've been told that there are 100 centimeters in a meter, they haven't internalized this concept.)

4.MD.A.2 Use the four operations to solve word problems involving distances, intervals of time, liquid volumes, masses of objects, and money, including problems involving simple fractions or decimals, and problems that require expressing measurements given in a larger unit in terms of a smaller unit. Represent measurement quantities using diagrams such as number line diagrams that feature a measurement scale.

Math Practices
1 Solve Problems & Persevere
7 Make Use of Structure

Sparkles' Mix-Up Mess!

This wacky tale about Sparkles the Clown, along with some good old-fashioned competition, provides students with a fast-paced and fun way to practice working with unlike units within the same system of measurement.

To prepare for this activity, copy and cut apart the Sparkles' Mix-Up Game Cards on page 223. You'll need 1 set of cards for each team of 3–4 kids. Students will also need pencils and a sheet of paper.

- Begin with a review of the relationship between inches, feet, and yards. Next, using your best dramatic flair, introduce Sparkles the Clown. Explain the chaos he has created in the school. Say, "Sparkles the Clown has made a mess in the school! He measured and recorded many things, but he goofed up each and every time!"

- Continue by saying, "For example, he said Zack was shorter than Zoe because Zoe is 4 feet and Zack is only 1 yard and 13 inches." Stop to discuss the absurdity of this clown! Say, "This is just one of his several mistakes. You'll work in teams to unravel the confusion that Sparkles created."

- Break kids into teams. Pass out a blank sheet of paper and Sparkles' Mix-Up Game Card #1, *face down*, to each team. With the cards still face down, give these instructions: "Teams, there are kids' names on the other side of this card, along with their heights. Your task is to order the kids from shortest to tallest and place this information on your paper in the form of a number line."

- Say, "But there's a problem. Sparkles mixed up the units of measurement, so it's going to be a little tricky. You must record your answers using the *same unit of measure* for every name. Work quietly so that other teams don't hear the answers from you."

- Say, "Raise your hand when your team has sorted all of the kids' heights. The first team to get all students in the correct order is the winner. Ready. Set. Go! Get busy and sort out Sparkles' mess."

- Cards flip fast and the fun begins! After the teams finish, discuss their answers. Did some teams choose inches while others compared in feet? Ask your students if this would make a difference. Compare the teams' answers and number lines.

- Repeat this routine for each of the Sparkles' Mix-Up Game cards. You may choose to do all 4 cards in one day, or you might prefer to use 1 card a day as a mental math warm-up or end-of-math session activity.

Here are the answers:

Card #1: Isa (30 inches or 2 feet, 6 inches), Colton (41 inches or 3 feet, 5 inches), Britta (45 inches or 3 feet, 9 inches), Vivian (48 inches or 4 feet), Dylan (49 inches or 4 feet, 1 inch), Wyatt (51 inches or 4 feet, 3 inches), Finn (54 inches or 4 feet, 6 inches)

Card #2: Cookie $0.41, Milk $0.45, Chocolate Milk $0.50, Fresh Fruit $0.95, Hotdog $1.25, Full Lunch $1.55

Card #3: Green food dye 0.003 liter, Yellow food dye 0.02 liter, Blue food dye 0.2 liter, Red food dye 0.43 liter, Seltzer 0.83 liter, Lemon juice 1.1 liters, Vinegar 2.023 liters, Rubbing alcohol 2.45 liters, Tap water 6.54 liters

Card #4: Crab Walk 4.2 meters, Sack Race 4.6 meters, Wheel Barrel 5.5 meters, Sneaker Relay 43 meters, Turkey Trot 48 meters, Wacky Water 50 meters, Grab Bag 52 meters, 3-Legged and Long Distance 91 meters, Flag Tag Relay 100 meters

Sparkles' Mix-Up Game Cards *Use with Sparkles' Mix-Up Mess!, page 166.*

Name: _____   Date: _____

Sparkles' Mix-Up Mess Card #1

OH NO! Sparkles got into the clinic and changed the heights on these children's charts. List each child's height from shortest to tallest using the same unit of measure. Show answers on a number line.

Colton: 41 inches
Dylan: 1 foot, 37 inches
Wyatt: 1 yard, 1 foot, 3 inches
Isa: $2\frac{1}{2}$ feet
Finn: $\frac{1}{2}$ yard and 36 inches
Britta: $1\frac{1}{4}$ yard
Vivian: 4 feet

Sparkles' Mix-Up Mess Card #2

OH NO! Sparkles posted these prices in the cafeteria. How confusing! List the price of each item in the cafeteria, using dollars and cents, from least to greatest. Show answers on a number line.

Milk: 4 dimes and a nickel
Chocolate Milk: 2 quarters
Cookie: 4 dimes and a penny
Hotdog: 125 cents
Fresh Fruit: 3 quarters, a dime, and 10 cents
Full Lunch: 5 quarters, 2 dimes, a nickel, and 5 pennies

Sparkles' Mix-Up Mess Card #3

OH NO! Sparkles mixed up the labels of these liquids in the science lab. Using the same unit of measure, list each liquid by amount from least to greatest. Show answers on a number line.

Lemon Juice: 1 liter and 100 milliliters
Vinegar: 2023 milliliters
Seltzer: $\frac{1}{2}$ liter and 330 milliliters
Tap Water: 6.54 liters
Rubbing Alcohol: 2450 milliliters
Red Food Dye: 430 milliliters
Yellow Food Dye: 0.02 liter
Green Food Dye: 3 milliliters
Blue Food Dye: 0.2 liter

Sparkles' Mix-Up Mess Card #4

OH NO! Sparkles changed the measurements for the Field Day Races. Using the same unit of measure, list each race in order from shortest to longest. Show answers on a number line.

Long Distance: 9,100 centimeters
3-Legged: 91 meters
Wheel Barrel: 550 centimeters
Sack Race: 460 centimeters
Crab Walk: 420 centimeters
Flag Tag Relay: 10,000 centimeters
Sneaker Relay: 43 meters
Wacky Water: 50.0 meters
Turkey Trot: 4,800 centimeters
Grab Bag: 5,200 centimeters

Kids have fun working together to fix Sparkles' mistakes.

 **QUICK TIP**

Pastry School in Paris: an Adventure in Capacity by Cindy Neuschwander is a delightful story that tells the relationship of measurements in story form.

4.MD.A.3 Apply the area and perimeter formulas for rectangles in real world and mathematical problems. *For example, find the width of a rectangular room given the area of the flooring and the length, by viewing the area formula as a multiplication equation with an unknown factor.*

Math Practices
1 Solve Problems & Persevere
7 Make Use of Structure

Storybook Character Problems

Kids need the freedom and time to figure things out! Present these problems just one at a time, over several days. Research shows us that working through 2 or 3 complex and multi-operational problems has a much more positive impact on student achievement than 30 or 40 drill-and-kill type problems (Leinwand, 2009).

Children should work on the first problem together in small groups, and then solve the last problem independently. Using storybook characters really helps to soften the blow of these toughies! To begin, post this problem:

Willie Wonka wants to create a rectangular chocolate bar with the largest possible area for a huge party. He has 2 choices. He can create a bar that's 13 feet long and 27 feet wide, or one that's 22 feet long and 18 feet wide. Which would give Willie the greater area?

➡ The 22 x 18-foot bar gives the greater area (396 square feet).

• Let kids work in small groups to discuss and solve. Move around the room and answer questions only if kids are at a standstill. Otherwise offer encouragement, suggest they use drawings, and ask them to double-check their computation.

• If necessary, ask prompting questions. Say, "What do you notice about the perimeters of these 2 chocolate bars?" (They're both the same: 80 feet.) Say, "If their perimeters are the same, then are their areas the same?" (No. No way! Give students a chance to remember or figure this out.) Ask, "How do we figure the area of the chocolate bar?"

• Insist that groups take turns explaining their answers, but don't tell anyone that they're right or wrong until every small group has spoken and there is consensus.

This next problem is multi-operational and relies on children understanding the dynamics of area and perimeter.

Mr. Al E. Gator and Miss Croco Dile want to build a rectangular fence. They have exactly 120 meters of fencing. They want the maximum amount of ooey-gooey swamp algae, boggy clumps, and swampy mist within their rectangular fence. Mr. Al E. Gator wants 1 side of the rectangle to be exactly 36 meters long. Miss Croco Dile wants 1 side of the rectangle to be exactly 34 meters long. Whose idea is correct? Describe your answer with illustrations, labels, and math models.

➥ Miss Croco Dile's plan is correct. It results in 884 square meters of lovely swamp muck.

- Introduce the problem. Say, "Boys and girls, here's a problem for you to solve on your own. Be a patient problem-solver, persevere, read and reread the problem, and then proceed one step at a time."

- If your students aren't ready to tackle this problem, then ask them to join those who want to discuss the problem for a short math meeting. At this meeting, go over the vocabulary and what the problem is asking them to do. Remind them how they solved similar problems with you before. Some children may need to refer to their notes on the similar problems.

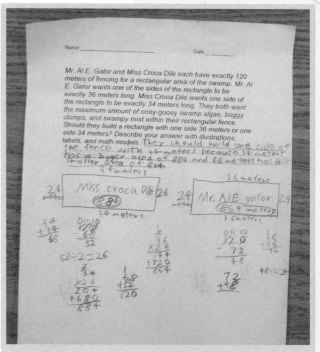

This child uses models, labels, and number sentences. His thought process and steps are clear and easy to read.

GRADE ④

Cluster 4.MD.B Represent and interpret data.

 Smithsonian Handbooks: Butterflies and Moths by David Carter

Wing It!

Kids will develop a deeper understanding of how to represent and interpret data on a line plot if you let them run with the kind of data that *they* find fascinating. Topics such as the lengths of various rodents, tropical fish, lizards, snakes, or creepy bugs are proven kid-friendly topics!

Since this standard requires using the fractional units $\frac{1}{2}$, $\frac{1}{4}$, and $\frac{1}{8}$, students will need standard rulers marked to eighths. They'll also need paper (notebook paper turned horizontally works well) and the butterfly and moth data provided in the chart on the following page.

- Curiosity is a wonderful way to get those eyes fixed on you, so use it! Ask, "Boys and girls, what would have a wingspan of 6 inches?"

Whole Group, Individuals

4.MD.B.4 Make a line plot to display a data set of measurements in fractions of a unit $(\frac{1}{2}, \frac{1}{4}, \frac{1}{8})$. Solve problems involving addition and subtraction of fractions by using information presented in line plots. *For example, from a line plot find and interpret the difference in length between the longest and shortest specimens in an insect collection.*

Math Practices
4 Model with Mathematics
5 Use Tools Strategically

Allow some predictions, and then post the butterfly and moth information provided below.

- Say, "Check it out, entomologists—this chart shows the wingspan of some moths and butterflies. A chart like this gives us lots of great information, but there's an even better way to show this data. It's called a line plot. A line plot not only gives us information, it gives us an easy way to make comparisons!"

- Continue by saying, "You're going to create a line plot showing the measurements of these different species of butterflies and moths. Once you've plotted the measurements and labeled the insect's length above the correct fraction or whole number, you can make lots of cool comparisons."

- Pass out paper and rulers. Instruct kids to follow your directions to create their own line plots. Say, "You'll draw a line. The line must have an arrow at each end. Mark your line to start at 0. Now look at the chart and find the butterfly or moth with the greatest wingspan. Yes, the Cecrophia Moth has the widest wingspan, so your line needs to be at least 6 inches long."

- Tell students, "Before we continue, we need to give our line plots a title. How about 'Butterfly and Moth Wingspans'?"

- Say, "The measurements provided for these insects on this chart are given in halves, fourths, and eighths. So we'll need to mark our

Butterfly or Moth	**Wingspan in Inches**
Citrus Swallowtail	$4\frac{1}{4}$
Giant Swallowtail	$5\frac{1}{2}$
Common-Dotted Border	$2\frac{1}{2}$
Long-Tailed Blue	$1\frac{1}{2}$
Paris Peacock	$5\frac{1}{4}$
Cassius Blue	$\frac{3}{4}$
Grass Jewel	$\frac{5}{8}$
Cecrophia Moth	6
Calleta Silkmoth	$4\frac{1}{2}$
Poplar Hawkmoth	$3\frac{1}{4}$

lines at every one eighth of an inch." Show students how to make a tick mark at each one-eighth interval (see photo).

- Next, go down the chart, 1 winged wonder at a time. Say, "Let's see. The Citrus Swallowtail is $4\frac{1}{4}$ inches. Find that spot on your line plot. Put an x directly above $4\frac{1}{4}$, which is also $4\frac{2}{8}$, and then write the word 'Citrus Swallowtail.' You'll want to turn your paper so that the word is written perpendicular to the line." Continue until all butterflies and moths are listed and marked on the line.

- Once the students' line plots are created, ask questions that use addition and subtraction and require kids to study the data on the line plot in order to answer.

- Pose questions such as: "What's the difference between the length of the Poplar Hawkmoth and Grass Jewel?" or "How much wider is the wingspan of the Paris Peacock than the Common-Dotted Border?" or "How much shorter is the wingspan of the Long-Tailed Blue than the Giant Swallowtail?"

- If your students are struggling, guide them to look at the closest whole number. For example, let's say you're comparing the length of the Grass Jewel $\left(\frac{5}{8}\right.$ of an inch$\right)$ to the Poplar Hawkmoth $\left(3\frac{1}{4}\right.$ inches$\right)$. Say, "Think. What measurement will get me to 1 inch from $\frac{5}{8}$? Yes, $\frac{3}{8}$. Now what fraction can get me from 1 inch to $3\frac{1}{4}$ inches? Yes, $2\frac{1}{4}$, which is also $2\frac{2}{8}$. Now, let's add $\frac{3}{8}$ to $2\frac{2}{8}$. The difference in length between the insects is $2\frac{5}{8}$ inches."

- Whenever you can, encourage kids to break the process into steps. Remind them that the first step is to get to the closest whole number and then add up. This strategy will serve them well in many situations, not just comparing wingspans!

Write About It: Pose the question: "How does a line plot help you to compare information?" Ask students to include a picture to support their ideas.

Your students must attend to precision as they make tick marks to show every one-eighth of an inch on their number lines.

C D A
Pairs

4.MD.C.5 Recognize angles as geometric shapes that are formed wherever two rays share a common endpoint, and understand concepts of angle measurement.

4.MD.C.5a An angle is measured with reference to a circle with its center at the common endpoint of the rays, by considering the fraction of the circular arc between the points where the two rays intersect the circle. An angle that turns through $\frac{1}{360}$ of a circle is called a "one-degree angle," and can be used to measure angles.

Math Practices
6 Attend to Precision
7 Make Use of Structure

GRADE **4**

Cluster 4.MD.C Geometric measurement: understand concepts of angle and measure angles.

Angles of a Crown

Hook your students with a Royal Challenge! As they work their way through this demanding problem, they'll make many brilliant discoveries. Give your mathematicians plenty of time to grapple with this one. Don't swoop in on the crowd. They need to persevere!

For this activity, students will work with a partner. Each pair needs a bag of pattern blocks (containing 1 yellow hexagon, 1 red trapezoid, 1 blue rhombus, 1 green triangle, 1 orange square, and 1 tan rhombus), paper, crayons, and pencils.

- Partner students and say, "The Pattern Block Queen has a magnificent crown made completely out of pattern block jewels. She's issued this challenge to her loyal subjects: Find the angle of every jewel on her crown."

- Continue by saying, "The Queen will give you 2 clues. Using these 2 clues you must unlock the measurement mystery for every piece in her crown." Pass out a bag of pattern blocks to each pair and say, "This bag contains 1 sample of each of her jewels."

- Say, "The first clue is for the orange square. We know this has 4 right angles. The Queen tells us the right angles measure 90°. The second clue is for the green equilateral triangle. This triangle has three acute angles. Acute angles measure less than 90°. Each of these acute angles measures 60°.

- Say, "From this point on you and your partner are on your own! Report to the group each time you find an unknown angle measure."

The chart on the following page shows some of the discoveries that your students may come up with. Keep a record of their findings on the whiteboard or a piece of chart paper.

Pattern Block Shape	Discovery	Reasoning
Yellow Hexagon	Obtuse angles = 120°	2 green triangle angles = 1 hexagon angle and 60° + 60° = 120°.
Red Trapezoid	Acute angles = 60°	The acute angles are congruent to the angles on the green triangle.
	Obtuse angles = 120°	The obtuse angles are congruent to the angles on the yellow hexagon.
Blue Rhombus	Acute angles = 60°	The acute angles are congruent to the angles on the green triangle.
	Obtuse angles = 120°	The obtuse angles are congruent to the angles on the yellow hexagon.
Tan Rhombus	Acute angles = 30°	It takes 2 acute angles to compose the angle on the green triangle. Since 30° + 30° = 60°, each acute angle of the tan rhombus is 30°. (Another way to say this would be that the 60° angle of the green triangle can be decomposed into two 30° angles.)
	Obtuse angles = 150°	The obtuse angle of the tan rhombus added to its acute 30° angle makes a straight line. 180° − 30° = 150°.

Figuring out the measurement of the tan rhombus's obtuse angle can be a bit tricky. You may need to remind students that there are 180° in a straight line. They can use 2 squares to compose a straight angle and find the answer, 90° + 90° = 180°. Once students know that a straight line is 180°, they can figure the measurement of the obtuse angle on the tan rhombus.

Extension: The Pattern Block Queen has one more challenge for her loyal subjects! She wants to know if any of the jewels in her crown can tessellate. (If an angle is a factor of 360—e.g., 30, 60, 90, 120—it will tessellate. For example, the tan rhombus will tessellate at its acute angle.)

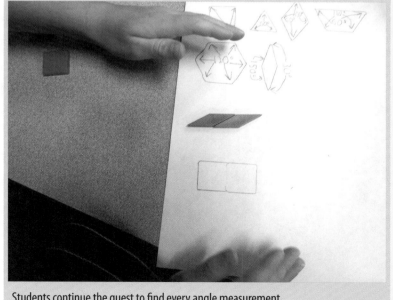

Students continue the quest to find every angle measurement.

4.MD.C.5a

**Mathematical
Practices**
4 Model with Mathematics
6 Attend to Precision

Angles in Plates

The angle ruler, also called a goniometer, helps students build a deep conceptual understanding of fractions. Once children understand how to use an angle ruler, the protractor will be easy to read!

For this activity, each student will need 1 set of Paper Plate Fractions, page 88, and an angle ruler. (Angle rulers are inexpensive and can be found online at www.hand2mind.com or in office supply stores.)

- If this is the first time your students are using the paper plate fractions, give them time to explore. They'll make many interesting and important discoveries about equivalence. Listen to their conversations and encourage everyone to share their discoveries with their peers.

- Ask each student to each pull out a $\frac{1}{4}$ paper plate fraction piece. Say, "Hold the angle ruler in the closed position so that you can read the writing on the ruler. Now open the top leg of the ruler and hug it to the angle of the $\frac{1}{4}$ fraction piece." Help students read the ruler to discover that the angle measures 90°.

- Say, "Since you know that the $\frac{1}{4}$ fraction piece has a 90° angle, can you find the angle measurements for the $\frac{1}{2}$ fraction piece and the $\frac{1}{8}$ fraction piece *without* using the angle ruler? You may turn and talk to your neighbor. Remember, you must justify your findings when you share them with the group."

- Students can manipulate the paper plate fraction pieces to see that it takes 2 quarters to make one half. Since one quarter has a measure of 90°, they can figure out 90° + 90° = 180°. This is a straight angle and students can see this concretely with the plates. Students can also see that two $\frac{1}{8}$ pieces are equivalent to a $\frac{1}{4}$ piece, so the measure of the $\frac{1}{8}$ piece is half of 90°, or 45°.

- Let the students share their discoveries on the document camera, or give them chart paper to record their findings. After students share their findings with the class, ask them to measure the $\frac{1}{2}$ piece and the $\frac{1}{8}$ piece with the angle ruler to check their work.

The one-fourth paper plate fraction has a right angle measuring 90 degrees.

- You may want to save this final question for a different day. Ask, "What's the total of all the angles when we use four $\frac{1}{4}$ pieces that each have a measurement of 90°? What about two $\frac{1}{2}$ pieces? Why does this happen?"

- Both demonstrate 180° + 180° = 360°. This is the essential understanding for this standard. An angle that turns through $\frac{1}{360}$ of a circle is called a 1-degree angle, and can be used to measure angles. Three-hundred-sixty 1-degree angles are needed to create an entire circle.

4.MD.C.5b An angle that turns through *n* one-degree angles is said to have an angle measure of *n* degrees.

Math Practices
4 Model with Mathematics
7 Make Use of Structure

Measuring Masterpieces

 Pablo Picasso: Breaking All the Rules (Smart About Art) by True Kelley

 Vincent Van Gogh: Sunflowers and Swirly Stars (Smart About Art) by Joan Holub

In this activity, students find the beauty of mathematics placed within great masterpieces. The titles listed above are only suggestions. There are so many wonderful books in this genre! Select your favorite child-friendly book on a famous artist.

Each student needs an angle ruler (see page 174), 3 colored pencils or markers, and a photocopy of a piece of artwork from the artist you choose to focus on. Please don't photocopy the artwork from the book! The National Gallery of Art (images.nga.gov) offers more than 29,000 copyright-free downloadable works of art that you can use.

- Begin by reading the story about the artist that you've chosen. Discuss the chronology of events in the artist's life and other important facts. (That way you're incorporating the CCSS Literacy standard RI.4.5 into your lesson too!)

- Say, "When you look at a piece of artwork you can learn something about the artist and the times he lived in, but did you know that you can also find mathematics in art? There are parallel and perpendicular lines, arcs, geometric shapes, and fractions involved in the subjects."

- Pass out photocopies of the artwork you've chosen. Say, "Let's look closely at this masterpiece. Do you see any angles in this painting? Are those angles acute or obtuse?" Let children discuss their findings with their classmates.

- Remind students of their work in Angles in Plates, page 174. Say, "Remember when you had to figure out the measurements of the angles in the paper plate fraction pieces? Today, we're going to focus on just the degrees of the arc, not the fraction."

- Tell students that using the angle ruler, they must find at least 10 different angles in the piece of artwork.

- Say, "Outline the 2 rays that meet at the vertex forming the angle with your colored pencils. Mark every right angle in one color,

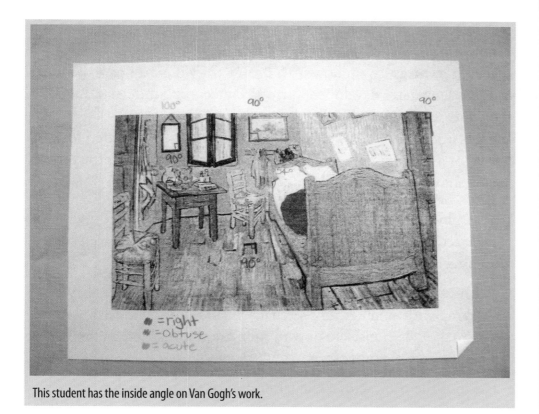

This student has the inside angle on Van Gogh's work.

every acute angle in another color, and every obtuse angle in a third color. Write the angle measurement near the angle. Be sure you write a key on your paper so I'll know what the colors represent."

• Once students are finished, project a copy of the artwork. Ask kids to describe the angles they measured. Encourage students to compare their measurements with each other. Needless to say, some kids may find the very same angle but with a slightly different degree of measure. This is the perfect time to bring out a compass and demonstrate how to measure accurately.

• To keep their skills sharp, offer different masterpieces throughout the year and have kids measure the angles and label the geometric shapes.

✓ **QUICK TIP**

Any of the books in the Getting to Know the World's Greatest Artists series by Mike Venezia are great for introducing your students to artists. We love the ones featuring Edward Hopper, Diego Rivera, and Grant Wood.

C ▶ P ▶ A

*Whole Group,
Individuals*

4.MD.C.6 Measure angles in whole-number degrees using a protractor. Sketch angles of specified measure.

Math Practices
1 Solve Problems & Persevere
3 Construct Arguments & Critique Reasoning
4 Model with Mathematics
6 Attend to Precision

Protractor Measurement

Learning to measure angles with a protractor will be a breeze because your students have developed a strong conceptual understanding of angle measurement using the angle ruler.

Each student will need a bag of pattern blocks (as prepared for Angles of a Crown, page 172), blank paper, rulers, and protractors. A document camera is helpful, but optional.

- Begin by saying, "Today you'll measure angles with a protractor. Protractors are a cool math tool. Rocket scientists, engineers, and architects use them!" Pass out the pattern block bags.

- With you modeling on the document camera (or, if necessary, working with small groups so that the children can see every move you make), demonstrate how to measure the angle of the orange square using the protractor.

- Pass out the protractors. Call attention to the tiny hash marks along the arc of the protractor. Ask everyone to count the number of tiny marks between the 2 rays that make the 90° angle. Say, "These represent the 1-degree angles we have been talking about. Did everyone count 90?"

- This is one tough concept so don't be surprised if you have to model and re-model this skill more than once. Ninety degrees is the easiest angle to measure because the 90° mark has only one number.

- When students are ready, say, "Measure one of the green triangle angles." This measurement has two choices: 120° and 60° (see photo).

- Ask students to turn and talk to a neighbor about which measurement is reasonable and why. Ask for a thumbs up or down for each choice. Discuss why 60° is the reasonable answer. (They are measuring an acute angle. Acute angles are less than 90°.)

- Follow this same procedure for the yellow hexagon, except give your students the opportunity to work alone. This will give you the chance to walk around and formatively assess each child while he's

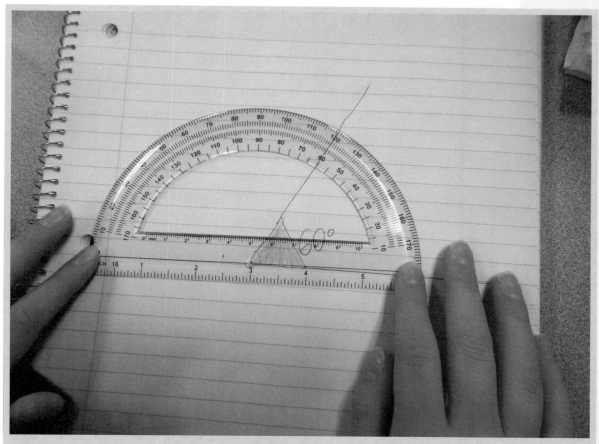

The green pattern block triangle links to protractor measurement. What choice is best? Is this angle acute or obtuse?

working. The measurement choices for the hexagon will also be 120° and 60°. Ask students, "Which choice is reasonable now?"

• This type of activity may be used for several days and then moved to centers.

 QUICK TIP

For detailed directions on how to teach using protractors, visit these websites:
www.mathplayground.com/measuringangles.html and
www.mathisfun.com/geometry/protractor-using.html

4.MD.C.7 Recognize angle measure as additive. When an angle is decomposed into non-overlapping parts, the angle measure of the whole is the sum of the angle measures of the parts. Solve addition and subtraction problems to find unknown angles on a diagram in real world and mathematical problems, e.g., by using an equation with a symbol for the unknown angle measure.

Math Practices
4 Model with Mathematics
6 Attend to Precision

Hamster Champs

 Hamster Champs by Stuart J. Murphy

Now that your students are measuring with protractors, it's time to ramp it up. In this activity students learn to see angles as additive. In other words, they learn how to add or subtract to find unknown angles. Students will need protractors, paper, and pencil.

- To begin, read the ridiculous and engaging story *Hamster Champs*. This fun-filled adventure is packed with racecars, ramps, and angles. The first ramp has an angle of 30°. That's not sufficient to make the racecar go fast enough for the daredevil hamster, Pipsqueak, or his friends. So they keep increasing the ramp's angle to add to the speed and flight distance of their car!

- After your students have enjoyed the story once, divide the class into small groups, pass out the protractors and paper, and have

Students measure 30 degrees and think, "How many degrees do I need to increase this ramp to reach 45 degrees?"

each group move to a table. Students will need plenty of room to measure and draw as they explore.

- Read the story again. This time ask children to draw what's happening with the ramp using protractors. The first angle measures 30°. This is followed by a 45° measurement. The third angle made by the hamsters measures 60°.

- After students have drawn and labeled the 3 different angles, ask them to write equations that show how many degrees the hamsters added to the ramp with each adjustment. Children may solve this problem using a missing addend ($30° + x = 45°$) or they may solve it using subtraction ($45° - 30° = x$).

- Tell students, "One idea the hamsters proposed was a 90° angle. How do you think this would work for a racecar ramp?" Let students argue this one out. (It wouldn't work because a 90°angle is straight up and down.)

- Ask, "What could you subtract from the 90° angle to make it work? Show your thinking with your protractor and a diagram." (As long as it's 89° or less it'll work.)

Extension: Let your kids reenact the hamster angles. Have them build a ramp with a 30° angle and race a toy car down the ramp. Next ask them to measure the distance the car rolls. Have them do this for the 45° and 60° ramps too. Ask students to compare and share their data.

GRADE ⑤

Cluster 5.MD.A Convert like measurement units within a given measurement system.

C P A
*Small Groups,
Individuals*

5.MD.A.1 Convert among different-sized standard measurement units within a given measurement system (e.g., convert 5 cm to 0.05 m), and use these conversions in solving multi-step, real world problems.

Math Practices
2 Reason Abstractly & Quantitatively
4 Model with Mathematics

The Ant's Progress

In this activity students listen to the tale of Digit the Ant and keep track of the distance Digit travels, using a meter stick and base-10 blocks. The base-10 blocks help give students a concrete way to understand the abstract movement of the decimal point in numbers. As the ant moves, students convert standard measurement units within the metric system.

You'll need to prepare a data sheet for each student as shown in the photo below. Every group of 2–4 students needs a meter stick and base-10 units and longs.

- Break students into small groups and begin the story: "Digit the Ant has found delicious picnic leftovers just 1 meter from his hill! As he travels to the picnic he wonders how far he has traveled."

- Pass out the meter sticks and base-10 blocks. Announce, "Every time Digit makes a stop, you'll work together to record his travels using the meter stick and the base-10 units and longs. Ready?"

- Say, "Start at 0. Digit's first stop is at 8 centimeters. He had to wipe some sand from his eye." Instruct the groups to place 8 unit cubes from the 0 to 8 centimeters along the meter stick. Add, "Digit also wants to keep track of how far he has traveled in meters, decimeters, and millimeters."

- Pass out the data sheets. Say, "Take a moment to record the measurement equivalences for 8 centimeters on your data collection recording sheet." (Students have been measuring with meters and centimeters since second grade. They should know that the meter stick has 100 centimeters.) Ask, "What fraction of the meter stick has the ant walked?" $\left(\frac{8}{100}\right)$ "How can you write this as a decimal?" (0.08)

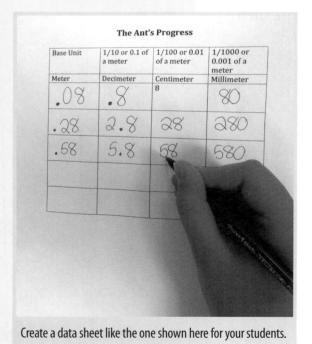

The Ant's Progress

Base Unit	1/10 or 0.1 of a meter	1/100 or 0.01 of a meter	1/1000 or 0.001 of a meter
Meter	Decimeter	Centimeter	Millimeter
.08	.8	8	80
.28	2.8	28	280
.68	5.8	68	580

Create a data sheet like the one shown here for your students.

Base Unit	$\frac{1}{10}$ or 0.1 of a meter	$\frac{1}{100}$ or 0.01 of a meter	$\frac{1}{1000}$ or 0.001 of a meter
Meters	Decimeters	Centimeters	Millimeters
0.08	0.8	8	80
0.28	2.8	28	280
0.58	5.8	58	580
0.80	8.0	80	800
1.00	10.0	100	1000

Filling out the data sheet provides students the opportunity to see that 1 meter = 10 decimeters = 100 centimeters = 1,000 millimeters. Knowing these benchmark metric measurements is essential to developing deep understanding.

- Students may require more help converting to millimeters and decimeters. Say, "Take a close look at the meter stick. Do you notice the tick marks? How many tick marks are between 0 and 1 centimeter? (10) These are called millimeters. There are 10 millimeters in each centimeter." Say, "Since each centimeter has 10 millimeters, how many millimeters has Digit walked?" (80) Have students record this information on their charts.

- Next discuss decimeters. Say, "Can someone tell me how many centimeters it takes to make 1 long? Yes, it takes 10, and 10 centimeters is called a decimeter. So, if 1 decimeter is equal to 10 centimeters, what fraction of a decimeter has Digit walked so far?" $\left(\frac{8}{10}\right)$ Ask, "How can this be written as a decimal?" (0.8)

- Continue with the story. Say, "Digit persevered longer before his next stop. He walked 20 centimeters before he stopped to rub one of his feet. How far has he walked now?" (28 centimeters) Make sure the students show this with the units and longs beside their meter sticks and record their answers on their data collection charts.

- Continue by saying, "Digit trudged on and managed to travel 3 decimeters before he stopped again. How many centimeters is 3 decimeters? (30 centimeters) What will the total be when we add the 3 decimeters to the 28 centimeters? (58 centimeters, 5.8 decimeters) How much of a meter would that be? (0.58) Has he walked more or less than half a meter? (more) Show this along your meter stick and continue to fill in the data collection chart."

Units and longs help students keep track of the ant's progress.

- Pause throughout the story to be sure that your students have included all stops. Help them notice that as they record centimeters, the decimeter and meter numbers are smaller because these are larger units.

- Say, "Digit trekked on; he really wanted that picnic lunch. Add 22 centimeters where he stopped to take a drink from a puddle." (The total is now 80 centimeters.)

- Conclude by saying, "Add 2 decimeters to his stroll and he arrives at the picnic!" (Total is 100 centimeters or 1 meter.)

GRADE ⑤

Cluster 5.MD.B Represent and interpret data.

C P A
Small Groups

5.MD.B.2 Make a line plot to display a data set of measurements in fractions of a unit $\left(\frac{1}{2}, \frac{1}{4}, \frac{1}{8}\right)$. Use operations on fractions for this grade to solve problems involving information presented in line plots. *For example, given different measurements of liquid in identical beakers, find the amount of liquid each beaker would contain if the total amount in all the beakers were redistributed equally.*

Math Practices
3 Construct Arguments & Critique Reasoning
5 Use Tools Strategically

Lemon Juice Line-Up

Your students have been working with line plots since they were second graders! Each year since, the mathematical demands on them have increased. Last year, as fourth graders, they looked at line plots and added and subtracted to solve problems. This year they must employ all 4 operations. These skills will serve students well far beyond the fifth grade. Much data in the "real world" is presented in precisely this way.

For this task you'll need a millimeter-sized eyedropper (usually available for free at your local pharmacy), and clean and empty 1- and 2-liter bottles. You'll also need to post the information in the chart shown on the next page. Students only need paper and pencil.

- To give students an understanding of relative size, first show your students the 1- and 2-liter bottles. Next, use the eyedropper to fill, drop-by-drop, a 1-liter bottle to 0.1 of a liter. It'll take 100 drops!

This will build suspense and create a benchmark for a tenth of a liter. Building benchmarks is critical.

• Now it's time to set the stage for the Lemon Juice Line-Up. Display the information in the chart below for your students.

Beaker	Amount of Lemon Juice	Beaker	Amount of Lemon Juice
#1	0.5 liter	#6	0.4 liter
#2	1.5 liters	#7	3 liters
#3	2 liters	#8	0.6 liter
#4	0.25 liter	#9	1.75 liters
#5	1.75 liters	#10	2.25 liters

• Say, "Lemon juice has been poured into 10 beakers in the science lab. Each beaker has a different amount of juice. Your job is to create a line plot on your paper to show this information."

• Continue by saying, "After you've created your line plot, you need to write at least 5 (or whatever number you think is reasonable for your students) mathematical facts.

• Say, "See if you can use all 4 operations, including how to determine how much lemon juice there would be if all of the juice were poured into 1 large pitcher and distributed evenly into the 10 beakers." (The total of all 10 beakers is 14 liters, easily divisible by 10 as 1.4 liters in each beaker.)

• Once the graphs are completed, compare your students' work. Since the data is the same, all graphs should look pretty similar. Capture the concept of this fact!

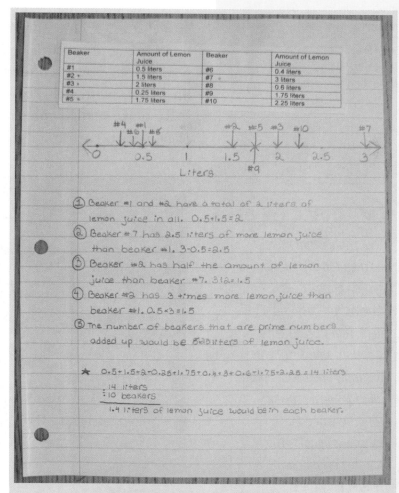

This student's careful listing and computation show solid understanding of the problem. This reveals so much more than a multiple choice test could!

Cluster 5.MD.C Geometric measurement: understand concepts of volume and relate volume to multiplication and to addition.

Building Cubes

In this hands-on activity students get to construct several cubes of different sizes. As your kiddos build their cubes, they're seeing and manipulating congruent faces, congruent edges, and congruent vertices. Those attributes are critical to determining the volume of a solid figure. Constructing a cube not only helps students deepen their understanding of what a cube is—it helps them visualize more complex shapes where all faces of a cube are not seen.

To prepare, you'll want to build a few cubes of different sizes as models to show to your students. In class, small groups will build duplicate copies of these cubes. You and your students will need poster or tagboard, coffee stirrers, straws, clay (marshmallows or spice drops also work well), tape, and scissors.

To make a small cube, cut 6 small congruent squares from poster or tagboard. Forming the paper squares into a cube is easy. Tape

Cubes constructed of coffee stirrers can be held in place with spice drops or clay.; toothpicks and marshmallows also work well.

This is what 1 square meter of children looks like.

4 squares in a vertical row. Next, being careful to match the sides exactly, tape the remaining 2 squares, one on the right of the second square and one on the left of the second square. Now fold those 6 squares into a cube. Tape the last 2 square faces in place and, voilà, you did it!

You can use coffee stirrers and clay to build a medium-sized cube. Connect 4 coffee stirrers using small globs of clay at the vertices/joints to create a square. Make a second square in the same way. Next, attach the two squares together using 4 more coffee stirrers to fashion a cube. To build an even larger cube, use straws that are longer than the coffee stirrers and hold them together with masking tape.

If you can get your hands on 8 dowel rods or plastic rods that are one meter in length and some modeling clay or small "3-way corner adapters" from the hardware store, then you and your students can create a gigantic 1-cubic-meter cube. Ask a few kids to climb inside—they'll all want to do this! This is guaranteed to create a visual of a cubic meter in their minds that'll last for a long time.

Write About It: Ask students to describe a cubic unit to someone who has never heard the term, using both words and labeled diagrams.

Small Groups

5.MD.C.3b A solid figure which can be packed without gaps or overlaps using *n* unit cubes is said to have a volume of *n* cubic units.

Math Practices
4 Model with Mathematics
8 Express Regularity in Repeated Reasoning

Counting Cubes

Very simply put, this standard wants students to understand that if they fill a solid figure with 6 cubes, and there's no space left and nothing overlaps, the volume of the solid figure is 6 cubic units. Short, sweet, and to the point. Notice that metric and customary units are not mentioned.

For this activity you'll need 1-centimeter and 1-inch unit cubes. You'll also need to make several rectangular prisms using poster or tagboard and tape.

Some of the rectangular prisms you make should have volumes in exact inches and some should have volumes in exact centimeters. It's important to use exact inches or centimeters so the cubes will fit perfectly inside each one with no gaps or spaces.

To construct the prisms use the same basic directions provided on page 186, but remember, when creating a rectangular prism, you need 3 pairs of congruent faces that are parallel to one another, as shown in the photo at the bottom of the page.

Leave one face open on each figure so that it can be opened and closed. This face will later be pulled back by students so they can

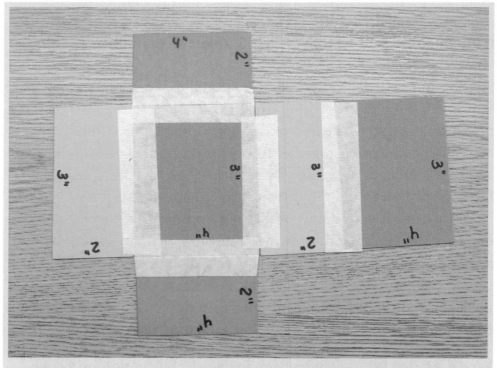

When creating a rectangular prism, you need 3 pairs of congruent faces that are parallel to one another. So to make a prism with a volume of 24 inches, for example, you'd have two 4 x 3-inch faces, two 3 x 2-inch faces, and two 4 x 2-inch faces.

place unit cubes inside. Label each figure you make using the letters, A, B, C, and so on.

You may want to mark the bottom faces of the metric-measured figures with a tiny "M" and the bottom faces of the customary-measured figures with a tiny "C." That way you'll know to place the 1-centimeter cubes with the metric-measured figures and the 1-inch cubes with the customary-measured figures.

- Say, "Okay, mathematicians, today we're going to fill different-sized rectangular prisms, which are also called 'figures,' with cubes. You're going to count the cubes as you place them in the figures. The number of cubes it takes to completely fill the figure, with no spaces or gaps and with nothing hanging over or towering above, is called the *volume*."

- Continue by saying, "The volume is the number of cubes that fit exactly inside the shape. So, if it takes 6 cubes to completely fill this figure, then you say its volume is 6 cubic units. The word *cubic* means you're measuring volume."

- Add, "Each of the figures you'll work with today has a different size. Some of the cubes are different sizes too. However, you'll only put like-sized cubes into a figure. Be sure you use the correct cubes. The cubes placed next to each figure are the ones that need to be counted and placed inside it."

- Don't name the unit size of the cubes (that comes in the next standard). Instead, refer to the centimeter cubes as, for example, "plastic cubes" and the 1-inch cubes as "wooden" or "foam" cubes. If you use sugar cubes, place them with the customary-measured figures. (The typical sugar cube is a $\frac{1}{2}$-inch cubic unit, but there's no need to share that information with your kids yet.)

- Give each small group a solid figure and at least enough cubes to fill the figure. (Remember, use 1-centimeter cubes with the metric-measured figures and 1-inch cubes with customary-measured figures.)

- Ask the children to fill each of their figures with the cubes and record their findings in their math journals. They can write simple

A variety of colorful prisms are filled with cubes that were counted as they were placed inside the prisms. The total number of cubes has been recorded.

statements such as, "Figure D is 16 cubic units."

- Either move the children to another center with a different solid figure and cubes, or move the materials from one group to the next. This way each group gets lots of experience counting and recording. Remember, solid conceptual understanding is built by repeatedly practicing the same concept using a variety of forms. (Merci, Monsieur Piaget, your wisdom does not grow old!)

Extension: Once students have experienced filling solid figures, allow them to create irregular solid shapes by using just the cubes (with no exterior container). While the shape doesn't look like a prism, it still has volume, and counting the cubes tells the volume!

C P A

Small Groups

5.MD.C.4 Measure volumes by counting unit cubes, using cubic cm, cubic in, cubic ft, and improvised units.

Math Practices
1 Solve Problems & Persevere
4 Model with Mathematics

Cubed and Then Counted

In the preceding standard, students were asked to count the cubes as they filled a figure. The final counted cubes gave the volume in cubic units. In this standard, the cubes are already placed in the figure and the children count those placed cubes to determine the volume. Another notable difference between these 2 standards is that now the size of the unit cube is named.

You'll need the solid figures and cubes used in Counting Cubes, page 188, for this activity, and students will need their math journals.

To prepare, fill the various solid figures with the appropriate cubic units. (Spend at least 1 day on metric volume and another day on customary volume. The order in which you do this doesn't matter.) You'll also want to build some irregular solid figures using just the cubes. The number of figures you prepare should be equal to or greater than the number of small groups of students.

- Say, "All right, my Einsteins, here's the task for today. You'll notice the solid figures we used before. This time they're already filled with cubes. You'll also see some irregular solid figures that I built." Hold up a 1-inch cube and say, "This cube is 1 inch wide, 1 inch high and 1 inch deep. It is called a 1-inch cubic unit." (Or, if you're using metric figures, hold up a 1-centimeter cube and give the cube's metric dimensions.)

- Continue by saying, "Each solid figure is filled with 1-inch (or 1-centimeter) cubes. Your task is to determine the volume of each solid figure. You may need to reach inside to find out how many cubes high the shape is; in other words, you may need to find out out how many layers there are since you can see only the top layer. It's that simple!"

- Say, "You'll each keep a record of your work in your math journal. For example, if this shape contained sixteen 1-inch cubes, then you'd write 'Shape B has a volume of 16 cubic inches.' We'll compare everyone's results once we're all done."

- First assign each small group a figure to begin with, and then allow children to move from one figure to the next.

- Challenge your students. Say, "While you're doing this activity, I want you to be thinking: Is there an easier, more efficient way to find volume other than counting all of the cubes?" (You're planting a seed here. Moving to the formula for finding volume is what the next standard is all about.)

Variation: After kids have gained experience with teacher-created figures, ask each group to create an irregular solid figure. Have the groups exchange their creations and determine the volume of their new figure. Let the original designers of each figure confirm the correct answer.

Extension: You can use sugar cubes for this activity, but if you do, the children will need to understand that it takes 4 sugar cubes to equal one 1-inch cube (the typical sugar cube is a $\frac{1}{2}$-inch cubic unit). This makes a great center for children who need more challenges.

Stacked sugar cubes are sweet way to show volume.

5.MD.C.5 Relate volume to the operations of multiplication and addition and solve real world and mathematical problems involving volume.
5.MD.C.5a Find the volume of a right rectangular prism with whole-number side lengths by packing it with unit cubes, and show that the volume is the same as would be found by multiplying the edge lengths, equivalently by multiplying the height by the area of the base. Represent threefold whole-number products as volumes, e.g., to represent the associative property of multiplication.

Math Practices
3 Construct Arguments & Critique Reasoning
4 Model with Mathematics

Cubes, Hoops & Smiles!

Get ready for some fun! Your students will be all eyes when you bring out the hula hoop and cubes. These kid-friendly tools help focus student attention on the learning. Children will move from the concrete to the abstract as they learn the formula for figuring volume. This activity also helps students experience the associative property.

You'll need 5 index cards, 1 hula hoop (from a dollar store or your P.E. teacher), and lots of wooden or plastic cubes. Each pair of students will need at least 60 cubes—if necessary, borrow extras from your friends who teach in the primary grades; they're guaranteed to have oodles of them.

Display this problem for your kids:

Emma is constructing a right rectangular prism with wooden cubes. She's giving her friend Judd clues about the prism. She tells him that the bottom layer is 3 cubes wide and 5 cubes long. It took 60 cubes to build the prism. She asks Judd, "How high is my prism?"
➡ Emma's prism is 4 cubes high.

* Partner students and provide them with cubes to build Emma's prism. Allow kids time to wrangle with this problem. When they look confident about their answer, ask, "How many cubes did the first layer take?" (15) Say, "If I know the bottom layer is 3 by 5, how could I calculate the number of cubes without counting?" ($3 \times 5 = 15$) Ask, "How many layers of 15 will you need to have a total of 60 cubes?" (4 layers)

* Tell students, "Judd told Emma that he doesn't even have to build the prism because he knows that $15 \times 4 = 60$. Who can tell me, where did Judd get the 15 from? (3×5) and what does the '4' represent in his equation?" (height)

* Help students connect to their prior learning. Say, "Remember how you filled all of those prisms with cubes in the last lesson? I asked you if there was a more efficient way to figure the volume rather than counting the total number of cubes. Well, who thinks they can tell me the equation for figuring the volume of this prism?" ($3 \times 5 \times 4$)

* Continue by asking, "What general statement can we use to find the volume of any rectangular prism? ($l \times w \times h = volume$) Say,

"When we write this equation we can label it this way: *l* × *w* × *h* = *volume*." As you write the equation, be sure the children understand that the *l* is for "length," the *w* for "width," and *h* is for "height."

- Ask, "Will this equation work every time?" (yes) "Does it matter how I group the factors?" (no)

- To prove that the order of the factors doesn't matter, take out a hula hoop and 5 index cards. Write "3" on one card, "5" on the next, and "4" on the third card. Make multiplication signs on the last 2 cards. Ask 5 student volunteers to hold the 5 cards.

These kids are proving the associative property to compute volume.

- Take the hula hoop and use it to circle the 3 kids with the cards labeled "3," "5," and "×." Say, "Judd grouped the factors like this. (3 × 5 = 15) So, now we take the product of 3 and 5, which is 15. Now multiply that by 4 and we get 60!" Look to the children with the other "×" and the "4" cards, but don't put the hoop over them! The hoop represents the grouping involved in the associative property.

- Next, move the hula hoop and circle the kids with the "5," "4," and "×" cards. Ask, "Will this work? Can I multiply 5 × 4?" Look to the 2 children with "×" and "3" cards and say, "Can I multiply 5 × 4 and then multiply that product times 3?" (Yes, 5 × 4 = 20 and 20 × 3 = 60.) Again, the hoop should only be used on the 5 × 4 here, representing the other possible grouping.

- Ask, "What mathematical property proves this will work?" (associative) While CCSSM does not expect our students to be able to name the properties, it does expect our students to know when to use the properties and to use them strategically.

You'll want to keep this hula hoop experience in your back pocket. It powerfully proves that the factors in multiplication can be grouped in different ways, but the product will not change.

Write About It: Show students a prism and say, "Write about how you can find the volume of this prism without building it." You may want to provide grid paper to students who are struggling. That way they can cut out the 6 faces as a visual representation.

5.MD.C.5b Apply the formulas $V = l \times w \times h$ and $V = b \times h$ for rectangular prisms to find volumes of right rectangular prisms with whole-number edge lengths in the context of solving real world and mathematical problems.

Math Practices
3 Construct Arguments & Critique Reasoning
4 Model with Mathematics
7 Make Use of Structure

Mummies and Math

 Mummy Math: An Adventure in Geometry by Cindy Neuschwander

In this story, Matt and Bibi are exploring the antiquities of Egypt with their famous scientist parents. In their quest to find a secret chamber, they encounter many geometric solids, including right rectangular prisms. In the antechamber of one ruin they find a clue on a piece of papyrus. It reads, "Look for a geometric solid with 6 identical faces."

In order for your students to embark on their own mathematical adventure they'll need cubes, grid paper, scissors, and their math journals.

After reading the story to your class say, "Each of you has a mission. Using cubes, you must design a geometric solid that fits the description of the clue in the story. You need to create a geometric solid with 6 identical faces." (Be sure to use the term "geometric solid" and not "cube," as that would give away the answer!)

Continue by saying, "After you do that, you'll need to write the formula that matches your replica in your math journal. Finally, you'll use grid paper to cut out a model of one of the faces from your

The largest cube students make is usually 5 x 5 x 5 in size because they run out of cubes. You can let this be a partner activity if you have limited numbers of cubes.

geometric solid. Hold on to your cutouts. We'll use them later in a whole-class discussion."

This assignment will produce many correct answers. Accept any size cube. Line up the cubes your students create in order from smallest to largest. Once all of the cubes are lined up, ask questions such as:

- How are all of these solids alike? (You hope they're all cubes!)
- What do you notice about the formulas/equations that go with these cubes? (All factors are the same.)
- Who made the smallest cube? (If no one made a 1 x 1 x 1, challenge someone to add this to the lineup. Continue this process until you have a 1 x 1 x 1 cube, a 2 x 2 x 2 cube, a 3 x 3 x 3 cube, and a 4 x 4 x 4 cube. Add more if time allows.)

Next, create a chart like the one shown here and record the dimension and volume for each cube. You may wish to create a large chart for the class to see and/or ask your students to create a chart in their individual math journals.

Dimensions	1 x 1 x 1	2 x 2 x 2	3 x 3 x 3	4 x 4 x 4	5 x 5 x 5
Volume	1 cubic unit	8 cubic units	27 cubic units	64 cubic units	125 cubic units

The next task is for students to work together to line up the graph paper faces they cut out in order from least to greatest. Say, "Each of these faces could be the base of one of the cubes." Continue by saying, "Let's calculate the perimeter and area for each face and add that information to our chart."

Dimensions	1 x 1 x 1	2 x 2 x 2	3 x 3 x 3	4 x 4 x 4	5 x 5 x 5
Volume	1 cubic unit	8 cubic units	27 cubic units	64 cubic units	125 cubic units
Perimeter of 1 face	4 units	8 units	12 units	16 units	20 units
Area of 1 face	1 square unit	4 square units	9 square units	16 square units	25 square units

Next ask questions such as:

- What do you notice about the pattern of the perimeters? (They are increasing by 4 units.) Why does this happen? (One example would be that the 1 x 1 square face has a perimeter of 4. When we increase this area to 2 x 2, we add 6 units of perimeter and

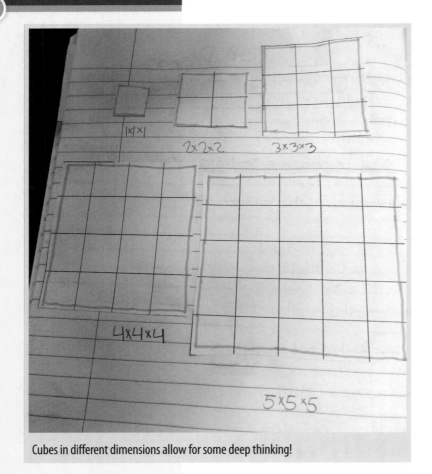

Cubes in different dimensions allow for some deep thinking!

cover 2 units for a gain of 4 units of perimeter.) The length of each side of each face of the figure is increasing by 1 unit. This repetition is done 4 times.

⊙ What do you notice about the pattern of the areas? (The increase is 3, 5, 7, and 9. These are consecutive odd numbers.) Kids need to be reminded of the patterns in the number system again and again, so don't pass over this opportunity!

Conclude this lesson by showing students the alternate formula for determining volume ($v = b \times h$). Say, "We've been using the formula ($l \times w \times h$) to determine volume, right? But let's look at our chart. If we multiply (the area of the base) × (the height) do we get the volume? Yes, we do!"

Break it down further. Say, "Since we calculate area by multiplying $l \times w$, we can multiply that product times the height and get the volume." Discuss why this works. You want students to understand that the reason multiplying the base × height gives volume is that in order to arrive at the base area, they've already multiplied the length times the width.

It's important for students to recognize both formulas for determining volume. Ask, "Who can state the 2 formulas we can use to find the volume of a right rectangular prism?" The 2 formulas stated in the standard need to be repeated. ($v = l \times w \times h$ and $v = b \times h$)

Write About It: Tell students that 1 face of a cube is 6 x 6. Say, "I want you to figure the volume of this cube and to show 2 ways you arrived at your answer." (6 × 6 × 6 = 216 or 36 × 6 = 216)

Pentominoes

 The Wright 3 by Blue Balliett

Pentominoes are the perfect tool for teaching this standard! Depending on your students, this hands-on investigation may take 2 days to complete.

Each child will need at least 5 tiles and at least 5 snap/interlocking cubes. (These are the cubes that can be added onto on all 6 faces. If you don't have them, check with your friends in the primary grades; those grades often have boxes of these great tools.) Students will also need their math journals and pencils.

For your reference, there are 12 possible pentomino shapes and they're named after letters of the alphabet. A picture of 2-D and 3-D pentominoes can be found in the front matter of *The Wright 3* book (they're also easily found online). But don't show these pictures to your students yet. It's much more fun if the kids discover the configurations on their own!

- Say, "Mathematicians, today you get to work on solving a fun and challenging puzzle. I'm going to give each of you 5 tiles. You must try to figure out all of the possible 2-D pentomino shapes. Each pentomino shape must use all 5 tiles." Explain to students that the tiles must be placed edge to edge so that they meet at the corners.

- Continue by saying, "After you discover a shape, make a simple sketch of it on your paper." As students discover and draw each figure, you'll want to assign the letter name to each pentomino.

- After all of the pentomino shapes have been created, named, and recorded, it's time for a quick review of perimeter and area.

- Ask, "Do all of these figures have the same area?" (yes) "Do they all have the same perimeter?" (no) "Why does the P-shaped pentomino have a smaller perimeter than

Pentominoes take right rectangular prisms to the next level.

> **5.MD.C.5c** Recognize volume as additive. Find volumes of solid figures composed of two non-overlapping right rectangular prisms by adding the volumes of the non-overlapping parts, applying this technique to solve real world problems.
>
> **Math Practices**
> **4** Model with Mathematics
> **7** Make Use of Structure

the L- or I-shaped pentomino?" (They should notice that the P-shaped pentomino conserves by having more interior sides touching.)

Once students have composed and discussed the twelve 2-D pentominoes, it's time to take this lesson to three dimensions!

- Bring out the snap/interlocking cubes and ask the children to make the 3-D version of the pentominoes. These can be made right on top of the 2-D versions. For example make an I-shaped pentomino by linking 5 cubes in a row.

- After all of the 3-D pentominoes have been constructed, ask, "What's the volume of each of your 3-D pentominoes?" (5 cubic units) Then ask, "Can you decompose the P-shaped pentomino into 2 right rectangular prisms?" (Yes, one is 2 x 2 x 1 and the other is 1 x 1 x 1.) "Does the volume of one of these parts plus the volume of the other part equal 5 cubic units?" (yes) "Why does this happen?" (Volume is additive; you can decompose the figure into parts and then add up the different parts/sections of the figure.)

- Continue this discussion by combining 2 pentominoes. Remember, you want your kiddos to be able to find volumes of solid figures composed of 2 right rectangular prisms by adding the volume of the 2 prisms. The L-shaped pentomino works well for this.

- Place 2 L-shaped pentominos side-by-side, and then link the cubes to form the new solid. Ask, "How can you decompose this new solid figure into 2 right rectangular prisms?" Allow students to discuss the 2 ways and ask the children to demonstrate this decomposition.

- Conclude this activity by bringing out the book *The Wright 3*. Read the first 12 pages. This delightful book is set in an area rich with Frank Lloyd Wright's highly geometric architecture, and the characters in the story use pentominoes on a daily basis!

 **QUICK TIP**

Hexominoes, polygons made of 6 squares, are used in a sample perimeter question on a CCSSM assessment.

Geometry

Children come to school with lots of geometric knowledge. Many have been stacking blocks, placing the circle shape inside the circle hole, and snapping puzzle pieces together since they were very young.

In the primary grades, students focus on the attributes of solid and plane shapes and they explore how they're alike and different. They also experience fractions through geometry. Intermediate grade children expand upon this knowledge immensely. Our intermediate students will also be hearing a new definition of trapezoid. A trapezoid is a quadrilateral that has at least 1 pair of parallel sides.

Third graders focus on plane figures as they study angles, categorize quadrilaterals, and work with fractions. Fourth graders learn to categorize triangles, and to draw points, angles, line segments, and rays. Coordinate grids make their debut in fifth grade, where students learn to correctly read and plot coordinate pairs and connect the points. Fifth graders are also expected to classify 2-D figures using a hierarchy of attributes.

Geometry is often regarded as the most beautiful area of mathematics, and it's truly everywhere our students look—ceiling and floor tiles, windowpanes, carpet patterns, sandwiches cut into triangles, balls they toss at recess.

The activities in this chapter capitalize on the "real-worldliness" of this branch of mathematics. Students will make quilts, work on construction teams to build triangles, and move around on a life-sized coordinate grid. Bringing the beauty of geometry to life in your classroom makes the learning powerful and fun!

Geometry is the only domain that crosses all grade levels, from kindergarten to eighth grade.

3.G.A.1 Understand that shapes in different categories (e.g., rhombuses, rectangles, and others) may share attributes (e.g., having four sides), and that the shared attributes can define a larger category (e.g., quadrilaterals). Recognize rhombuses, rectangles, and squares as examples of quadrilaterals, and draw examples of quadrilaterals that do not belong to any of these subcategories.

Math Practices
4 Model with Mathematics
7 Make Use of Structure

GRADE ③

Cluster 3.G.A Reason with shapes and their attributes.

Quadrilaterals

 If You Were a Quadrilateral by Molly Blaisdell

The animals in this whimsical story love drawing all sorts of 4-sided figures—kites, checkerboards, yoga mats. It's the perfect book for introducing your students to quadrilaterals.

For this activity students need geoboards, geobands, and a sheet of geoboard dot paper, which can easily be found online for free. If you don't have geoboards, students can just use the dot paper.

To prepare, create 4 paper signs: "Rectangle," "Square/Rectangle," "Rhombus," and "Not a Rhombus or Rectangle."

- Read *If You Were a Quadrilateral* to your kids, stopping to discuss any new math vocabulary. Note that page 20 of this story contains an older definition of the term "trapezoid."

- Tell students that according to the most current definition, a trapezoid is a quadrilateral with at least 1 pair of parallel sides (rather than exactly 1 pair of parallel sides), which means parallelograms are now also considered trapezoids. (For more information on this new definition, see Sort It Out, page 214.)

- Pass out the geoboards and geobands and say, "Please create a quadrilateral on your geoboard. Remember, quadrilaterals have 4 sides and 4 angles, and all sides must be straight."

- When they're finished, tape the signs you made in 4 different areas of the room and ask your students to get up and place their geoboards near the correct sign.

- Say, "Let's check out the quadrilaterals you placed in the rectangle area. Do they fit the definition on pages 12 and 13 of the book?" Study each example. If one is incorrect, let the group decide why it doesn't belong. Ask a student to place that geoboard in the correct area.

- Continue by saying, "Page 14 says that a square must have all equal sides and angles. Let's check out the geoboards placed in the square area." Be sure to discuss the fact that a square is a rectangle!

- The rhombi are next. Say, "The Rowdy Rhombuses on pages 16 and 17 tell us that a rhombus must have 4 equal sides." Validate the shapes placed in that area. A question that often comes into play during this discussion is, "Is a square a rhombus?" Ask, "Does a square have 4 sides that are equal?" (Yes and yes; a square is a rhombus with right angles.)

Is a square a rhombus or is a rhombus a square?

- Parallelograms aren't specifically mentioned in the standard, but the square, rectangle, and rhombus all fall into this category. Pages 18 and 19 zoom in on this shape. Ask, "Could you make a parallelogram that's not a square, rectangle, or rhombus?" (yes)

- The last category is "Not a Rhombus or Rectangle." Discuss the attributes of the geoboards that have been placed near that sign. Two examples of these shapes are trapezoids and 4-sided polygons. One type of trapezoid that students are familiar with is the red trapezoid block from the pattern block set.

Write About It: Pass out the dot paper and ask students to draw and label different types of quadrilaterals.

Extension: Challenge your crowd to create a 4-sided polygon that doesn't fit into any of the categories discussed in the lesson. (This shape will have no equal sides or 1 or 2 sets of parallel sides.)

Whole Group

3.G.A.2 Partition shapes into parts with equal areas. Express the area of each part as a unit fraction of the whole. *For example, partition a shape into 4 parts with equal area, and describe the area of each part as $\frac{1}{4}$ of the area of the shape.*

Math Practices
1 Solve Problems & Persevere
3 Construct Arguments & Critique Reasoning

Paper Quilt

 The Keeping Quilt by Patricia Polacco

 Sweet Clara and the Freedom Quilt by Deborah Hopkinson

The stories suggested above are just two gems from a rich collection of stories about quilts. Whatever title you choose for inspiration, your class's quilt will beautifully piece together geometry, fractions, and measuring!

Each student needs one 9-inch square and four 4-inch squares cut from construction paper. (Offering 6 different color choices for the smaller squares adds great variety to the finished class product.) Students also need markers, rulers, and glue.

- Pass out the large squares and rulers. Capture this time to review/ introduce some enriching geometric vocabulary terms. It'll only take a few minutes, and it's so easy to do!

- Begin by saying, "Boys and girls, let's talk about this shape. Yes, it's a square. It has 4 right angles and all of its sides are equal lengths. Yes, it's also a rectangle. It has 4 right angles. And it's a parallelogram, too, because the opposite sides are parallel! Is it a rhombus? Of course! Every square is a rhombus because all 4 sides are equal."

- Say, "The first thing you need to do is divide your square into 4 parts; each part must be an exact square. Any ideas how we can do this?" Let them visualize this; it's good for them.

- Confirm and demonstrate. "Yes, make 1 vertical fold, and 1 horizontal fold. Using your ruler as a guide, draw a straight line on both fold lines. You'll have 4 squares. Each one is $\frac{1}{4}$ of the whole."

- Point as you talk, "Let's call this 'Square 1' (top left), this 'Square 2' (top right), this 'Square 3' (bottom left), and this 'Square 4' (bottom right)."

- Let children select 4 smaller paper squares, each a different color. Instruct students how to cut their smaller squares. When they're ready to glue, have them leave space between the shapes so the background color will show through.

Mount the students' pieces close together so that they look like one huge quilt. Display their writing samples nearby.

Square 1: Two Rectangles

- ⊙ Fold the square in half so opposite sides line up.
- ⊙ Open it and cut on the fold.
- ⊙ Swap 1 rectangle with a friend who used a different color paper.
- ⊙ Glue your 2 rectangles in Square 1 vertically or horizontally.

Square 2: Four Triangles

- ⊙ Fold the square so opposite right angles match. You'll have triangles.
- ⊙ Open the paper and then fold it again so the other opposite right angles match.
- ⊙ Cut the 4 triangles. Swap 2 with 2 friends who used a different color paper.
- ⊙ Glue your 4 triangles in Square 2.

Square 3: Eight Triangles

- ⊙ (Follow the first two steps above for Square 2.)
- ⊙ Open up the paper and fold it so the opposites sides match, making a vertical fold.
- ⊙ Open it again and make a horizontal fold. You've made right triangles.
- ⊙ Cut out the 8 triangles. Swap 4 with 1 friend who used a different color paper.
- ⊙ Glue your 8 triangles in Square 3.

Square 4: Six Rectangles

- ⊙ Gently roll the paper, dividing it into thirds. Once you're fairly close to $\frac{1}{3}$, press hard on the folds.
- ⊙ Open the paper and then fold it in half, in the opposite direction of the $\frac{1}{3}$ fold lines.
- ⊙ Cut the 6 rectangles. Swap 3 of them with 1 friend who used a different color paper.
- ⊙ Glue your 6 rectangles horizontally or vertically in Square 4.

Wrap up the lesson by asking students to write about the 4 quarters of their quilt squares in their math logs. Require them to describe the shapes and to use fractions. (You may want to ask children to copy their teacher-edited mathematical observations onto index cards and mount those near the quilts.)

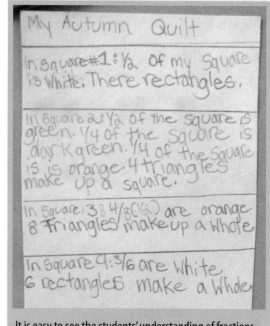

It is easy to see the students' understanding of fractions and geometry.

GRADE ④

Cluster 4.G.A Draw and identify lines and angles, and classify shapes by properties of their lines and angles.

Construction Site

 The Greedy Triangle by Marilyn Burns

The Greedy Triangle is oozing with mathematical conversation starters and instruction. When reading the book, encourage kids to look closely at each page. The illustrator has brilliantly placed a variety of geometric shapes on every page!

Divide your class into 5 small "construction crews." Each crew needs its own unique set of building supplies such as geoboards and geobands, a 4- or 5-foot-long jump rope or piece of string, a 6- to 8-foot piece of ribbon, a box of toothpicks and half a bag of gum-drops, or a box of straws and the other half bag of gumdrops.

Crew members also need paper, rulers, and pencils to document the structure their team builds. You absolutely must require drawings! That's at the heart of this standard.

This construction crew is busy at work.

- With children seated close enough to see the pictures, read the first 2 pages of the story. Direct your students' attention to the many different types of triangles illustrated.

- Ask questions such as: "Can you find an equilateral, acute triangle?" (The music triangle.) "Do you see the obtuse triangle?" (The pink house rooftop.) "Can you find perpendicular lines in the triangles or other places on these pages?" (The right triangles and the tablecloth.)

- Use vocabulary terms such as "line," "line segment," "ray," "angle" (right, acute, obtuse), "perpendicular," and "parallel" with your students. Expect them to use the correct terms from the get-go, and correct them if they confuse terms.

- Now it's time for some hands-on learning. Station your construction teams around the room. Instruct each crew to construct 1 triangle with their supplies.

- Once construction is complete, move around the classroom and discuss each team's triangle. (There should be a variety of triangles.) Discuss the attributes of each shape.

- After you've approved each crew's construction, require the crew to make a "blueprint" of their shape. Insist they use rulers or straightedges and attend to precision. Require them to identify and label the terms of their 2-D shapes as shown in the photo.

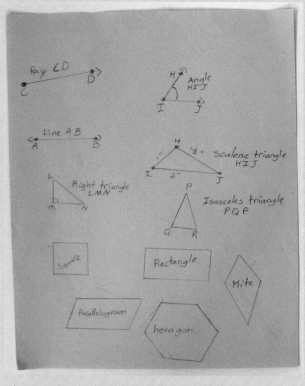

Blueprints show understanding and great vocabulary.

As the story continues, there are several more polygons that could be constructed and drawn. You may want to extend the construction to a second day or more. Splitting this lesson into 2 or 3 shorter sessions allows you to do other activities in your math lesson while giving kids repeated practice with the same geometric standard. Plus there's the excitement of returning to the construction site!

Extension: After your students have read, built, and drawn the shapes in this book, they may enjoy turning this story into a play. Visit: www.mathsolutions.com/documents/greedy_triangle_play.pdf for a script of *The Greedy Triangle*!

 QUICK TIP

You can find easy-to-follow, student-friendly definitions of the terms used in this lesson and more on the Math Is Fun website at: www.mathisfun.com/definitions/index.html.

***Whole Group,
Individuals***

4.G.A.2 Classify two-dimensional figures based on the presence or absence of parallel or perpendicular lines, or the presence or absence of angles of a specified size. Recognize right triangles as a category, and identify right triangles.

Math Practices
1 Solve Problems & Persevere
6 Attend to Precision

Peacock Power!

You'll begin this activity by showing students how to make a peacock that's close to 8 feet tall! Afterward, they'll construct their own paper birds using the same dimensions as yours with one big difference— their shapes will be measured in inches.

You'll need three 7-foot pieces of string or yarn, two 12-inch rulers and 2 yardsticks (or eight 12-inch rulers), and a marker or pencil for your demonstration. You'll also need plenty of floor space.

Students will need a copy of the Make a Peacock copymaster on page 224, construction paper scraps in a variety of colors, scissors, rulers, a compass or circle template, and glue.

- Gather kids around you. Announce that you're going to construct an enormous peacock from geometric shapes. Say, "I want to create its body from an isosceles triangle that has a 2-foot-long base and is 6 feet high. How could I measure a triangle that size using these tools?"

- Show students the string and measuring tools. Pause; their brains are warming up. After some wait time say, "Right, we can lay down two 12-inch rulers to make a base that's 2 feet long."

- Continue by saying, "I want the triangle to be 6 feet tall and the other two sides of the triangle must be the same length. What could we do? (Wait. The wheels are spinning.) Yes, putting the 2 yardsticks together makes 6 feet! Let's place them on the floor starting from the very middle of the 2-foot base."

- After your mathematicians figure out where the middle is, have a volunteer arrange the yardsticks on the floor. Ask, "How can we make the other 2 sides of the triangle? Bingo! We can connect the points from the top of the yardstick where we measured 6 feet to both ends of the base using yarn."

- Pull out 2 of the pieces of yarn. When you place the yarn on the floor, the students will see they reach above the apex/top of the triangle. You'll need to cut the extra from both pieces of yarn.

A floor-sized model of a peacock gives kids a great start.

- Ask, "How can we know if the 2 sides are equal?" Place the lengths of yarn side-by-side to prove they're equal. (You'll want to pick up all the rulers and yardsticks for the next step.)

- Say, "I want all of the tail feathers to be right triangles. Remember, a right triangle has a right angle and the 2 lines that form the right angle are perpendicular. How could we create a right triangle that's 2 feet long and 5 feet high?"

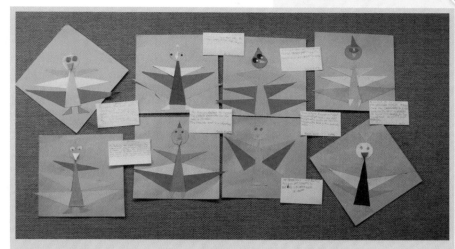

These student-made proud peacocks are smaller versions of the floor-sized model.

- Kids will have suggestions. You hope they'll decide that putting 2 rulers on the floor for one side and then 5 rulers at a right angle to represent a 5-foot height works perfectly.

- Use a piece of string or yarn to demonstrate that third side. No measuring needed. Tell kids that they'll use their rulers to draw a straight line for this side of the triangle that mathematicians call the "hypotenuse" (the side opposite the right angle).

- Say, "The other shape this peacock needs is a circle for its head. How could we draw a circle with a diameter of 2 feet using these materials?" Remind the children that all points on a circle are equidistant from the center, so the circle will have a radius that measures half its diameter.

- You may want to create a primitive compass with a string tied to a marker, leaving a 12-inch length from marker to thumb. Swing that marker around in a complete circle.

- After you've demonstrated how to make these 3 basic shapes, students should be ready to venture out on their own. Say, "You'll need to use these same techniques to create your own paper peacocks, but your birds' measurements will be in inches, so they'll be much smaller. Also, you'll use a compass to make the circles needed for your peacock's head."

- Pass out the materials and the Make a Peacock copymaster. Say, "I'll check your measurements for exactness. Go step-by-step and

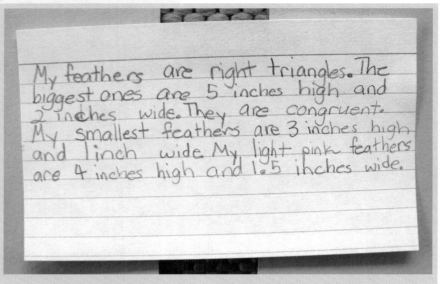

My feathers are right triangles. The biggest ones are 5 inches high and 2 inches wide. They are congruent. My smallest feathers are 3 inches high and 1 inch wide. My light pink feathers are 4 inches high and 1.5 inches wide.

Never miss an opportunity for kids to write about their thinking. It's your window to what's going on inside those beautiful minds.

check off each step as you complete it."

- Once the birds are complete, you'll have a colorful and proud display of geometry and measurement skills.

Write About It: Ask students to describe their handsome peacocks using beautiful mathematical language.

C P A

Whole Group

4.G.A.3 Recognize a line of symmetry for a two-dimensional figure as a line across the figure such that the figure can be folded along the line into matching parts. Identify line-symmetric figures and draw lines of symmetry.

Math Practices
1 Solve Problems & Persevere
3 Construct Arguments & Critique Reasoning

Hawaiian Quilt

Symmetry has traditionally been taught in all primary grades. But with CCSSM, fourth grade is now the first time this concept shows up. In this lesson, students construct a symmetrical paper quilt inspired by our beautiful 50th state.

Hawaiian quilts demonstrate the beauty of symmetry in an unforgettable way. Legend says shadows cast from the branches from the breadfruit tree were the early inspiration for these works of art. Hawaiian quilts were not created to add warmth on chilly nights; they were sewn for their beauty.

Students will need scissors, two 8-inch squares of construction paper (each a different color), and glue. You'll want to project images of Hawaiian quilts (visit www.hawaiianquiltartist.com). Most quilts use just 2 colors and employ rotational symmetry. Be sure to point out how these quilts are similar to an 8-sided snowflake.

- After an introduction to the quilts, pass out the two 8-inch paper squares to each student and say, "Leave 1 square on your desk. That will be your background color. Pick up the other square and fold it in half vertically. Press hard on the fold. Don't open it. Now it fold again, this time horizontally. Press hard. You'll have a small square."

- Check that everyone has a nice, flat square. It's important that the folds are neat and smooth. Say, "Place your square on the table so that the fold at the very center of your square is on the top left. Fold your square one last time so that the top-right right angle meets the bottom left right angle. You'll have a triangle."

- Say, "Now you'll make one cut. Begin your cut at the bottom part of the triangle and 'wiggle' cut from the shortest side of the triangle up to *almost* the center. Leave part of the center and cut toward the outside (the left side) of your triangle."

- Ask, "Can you visualize the shape you'll see when you open your paper? Your square has been folded into eighths. Open 1 fold and take a look. Do you see that fold right down the middle? That's the line of symmetry. Your shapes on both sides of that fold should be exactly alike."

- Say, "Now open your cutting 1 more fold." There will be oohs and aahs. "Check out your shape. Do you see the fold in the very middle? That's another line of symmetry. Imagine what you'll see when you open up your paper the whole way."

- Let students completely open up their papers. Ask, "What do you see? A beautiful shape! How many lines of symmetry do you see? (4) Check out how each side of the line of symmetry is exactly like the other side."

- Say, "Glue your cut-out shape to your background square. Glue your leftover parts, forming a negative image of your shape, to the other side."

- To display, tape everyone's quilt squares together. If you hang the final product in front of a window, there will be two slightly different views of this gorgeous, symmetrically cut quilt!

The finale is this lovely paper Hawaiian quilt.

The reverse side of the quilt is beautiful too!

Whole Group

5.G.A.1 Use a pair of perpendicular number lines, called axes, to define a coordinate system, with the intersection of the lines (the origin) arranged to coincide with the 0 on each line and a given point in the plane located by using an ordered pair of numbers, called its coordinates. Understand that the first number indicates how far to travel from the origin in the direction of one axis, and the second number indicates how far to travel in the direction of the second axis, with the convention that the names of the two axes and the coordinates correspond (e.g., *x*-axis and *x*-coordinate, *y*-axis and *y*-coordinate).

Math Practices
4 Model with Mathematics
6 Attend to Precision

GRADE ⑤

Cluster 5.G.A Graph points on the coordinate plane to solve real-world and mathematical problems.

Move It!

 A Fly on the Ceiling by Julie Glass

This humorous, tongue-in-cheek story about the life of René Descartes describes how he created the Cartesian coordinate system. After you set the scene with this book, your students have the chance to get up and move around on a life-sized coordinate grid!

You'll need a roll of blue or green painter's tape (it won't leave sticky goo on the carpet or tile) and index cards. Students will need pattern blocks and paper grids. (We make 12 x 12 grids created in Word and don't include any labels or numbers—we like our kids to label the grid themselves.)

To prepare, use painter's tape and make grid lines on your classroom floor. A 12-inch-square tile floor makes this very easy to do. Write the terms "*x*-axis," "*y*-axis," and "origin" on index cards and affix the cards to your grid. You'll want to write numbers on index cards and tape them to the grid too. If you make a grid that's at least 4 x 4, you'll have hit all of the important concepts.

- Your manipulatives are the kids themselves. Have them stand around the perimeter of the grid as you give directions. Ask one student to move to the origin (0,0).

- Say something like, "Adam, please move from the origin to coordinate pair (2, 4). That means you're going to move 2 spaces on the *x*-axis, which is also 2 spaces away from the *y*-axis. Next you'll move up 4 spaces away from the *x*-axis." After the student moves, write this coordinate point on the board.

- There will be wiggles and shuffling feet as the kids hope to be called on next. Continue, "All right, stay there, Adam. Now Meaghan, move to the origin and then move to coordinate pair (4, 2)." Point out how both children had a 4 and a 2 in their pair but landed in different spots.

- Students love this activity and will be happy to spend lots more time moving on the grid. You may have to help guide some children from one place to another.

- After lots of this concrete practice, pass out the 12 x 12 paper grids. Talk through each element of the coordinate grid. Have kids label the "*x*-axis," "*y*-axis," "origin," and the numbers.

- Next pass out the pattern blocks. Instead of simply saying, "Place the red trapezoid on the coordinate (3, 6)," you're going to take things up a few notches.

- Say, "Find this pattern block: It's a 4-sided polygon with 1 pair of parallel sides and 1 pair of non-parallel sides. It has 2 acute angles and 2 obtuse angles. All of its angles total 360 degrees." Scan the crowd; you hope everyone has the red trapezoid. Say, "Place *that* shape at (3, 6)."

The kid-sized grid lets students step into the math!

- Continue in this same manner. Give a clue that describes a pattern block shape and then ask students to place that shape on a specific ordered pair of coordinates. After giving 2 or 3 examples, let kids take turns giving the geometric clues and coordinates.

- Once the class understands the procedure, they can split into small groups and take turns giving their group members clues. This allows more kids the chance to use the language of mathematics.

Write About It: Ask students to explain the coordinate system to someone who has never heard of it, highlighting all mathematical terms in yellow.

In this activity students learn to use a coordinate grid *and* they brush up on geometry definitions!

C P A

Whole Group

5.G.A.2 Represent real world and mathematical problems by graphing points in the first quadrant of the coordinate plane, and interpret coordinate values of points in the context of the situation.

Math Practices
1 Solve Problems & Persevere
5 Use Tools Strategically

Coyote's Night Sky

 Coyote Places the Stars by Harriet Peck Taylor

After reading this Wasco Native American legend about the constellations, your student star gazers plot points and draw line segments to record constellations in Coyote's night sky on a coordinate grid. The final product for each child is 3 animals plotted on the coordinate grid, along with a set of ordered pairs/directions for each animal.

Each child will need a sheet of 2-centimeter grid paper, a ruler or straightedge, 1 sheet of notebook paper, and a pencil. This lesson may extend over 2 days.

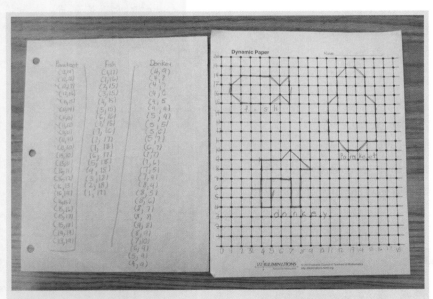

Students create their own animal-shaped constellations and the ordered pairs necessary to recreate each one.

- Pass out a sheet of grid paper to each student, and display an identical grid for students to reference. Review the terms "origin," "*x*-axis," "*y*-axis," and "coordinates." Discuss ordered pairs, the use of parentheses, how points are plotted, and how line segments can connect points.

- Say, "Your task is to plot coordinates on this grid so that when you connect the coordinates with line segments using a ruler, you'll get a shape that looks like a constellation in Coyote's sky." Of course, you're not going to get an exact owl or other animal, but do your best."

- Say, "I'll demonstrate how to draw a simple rectangle. First I'll plot a point at (4, 2), and then I'll place one at (4, 5). Next I'll plot a point at (2, 5). My last point will be at (2, 2). Now I'll connect the points with line segments using a ruler, in the same order that I plotted each one." Connect the points and say, "If I connected them in a different order, I might not have made a rectangle."

- Continue by saying, "You'll each make 3 animal constellations. Once 1 critter is on your grid, you'll need to write the coordinates for it in a list on the notebook paper. Your list must be written in the same order that you connected the coordinates with line segments. That way someone could follow just your ordered pairs and be able to create an animal exactly like the one that you created."

- Remind students that since they're making 3 animals, they'll need to make 3 coordinate lists.

A novel way for students to check work is to stack their grids together and hold them up to the light to see if they match.

Not too many fifth grade teachers are saying," Gee I wish I had more papers to grade," and checking the correctness of the ordered pairs could be a real pain in the neck. But never fear—here's a super-easy way to check the work that gives students more practice!

- Collect everyone's original coordinate grids with the constellations drawn and place them aside. Gather everyone's direction sheets with the ordered pairs listed. Next, making sure no one gets the original directions they wrote, pass out the directions, along with a blank coordinate grid, to students. Ask students to follow the ordered pairs and draw each animal constellation on the grid.

- If the ordered pairs were written and followed correctly, both grids will look identical. Simply have students stack both papers and hold them up to the light. If the constellations don't match, it's time for the 2 star gazers to conference. It's wonderful if your students can identify the problem and figure out how it can be corrected. Sometimes, a third student or you may need to join in to help sort things out.

 QUICK TIP

If your students need more practice with plotting coordinate grids, you may want to try Patterns with Rules on page 47.

Cluster 5.G.B Classify two-dimensional figures into categories based on their properties.

C ▶ P ▶ A
Whole Group

5.G.B.3 Understand that attributes belonging to a category of two-dimensional figures also belong to all subcategories of that category. *For example, all rectangles have four right angles and squares are rectangles, so all squares have four right angles.*

Math Practices

3 Construct Arguments & Critique Reasoning
6 Attend to Precision
7 Make Use of Structure

Sort It Out

In this activity students create a great visual, via a wall or bulletin board, of the categories and sub-categories used to describe triangles. This isn't a one-shot deal! Plan at least 2 days for this lesson.

You'll need index cards and a marker. Your kids will need paper (approximately 4 x 6 inches), rulers or straightedges, and pencils.

Yes, your students were taught about triangles before coming to fifth grade. However, if their understanding of the attributes used to describe these shapes is rusty, a stop at the website mathisfun.com/geometry would be time well spent.

- Pass out the paper and instruct kids to draw a triangle, any triangle. The greater the variety, the better! (Make an equilateral triangle and keep it handy in case no one else makes this kind.)

- Gather kids close to you on the floor, or move to a large table in the classroom. If possible, move to an empty cafeteria so there's a large amount of table space. Write the label "Triangles" on an index card. Place it on the floor or table.

- Tell children to place their shapes below the label. After a quick discussion about why the shapes are all triangles, say, "Let's sort them by the lengths of their sides."

- Write two new labels, "Scalene" (all sides are different lengths) and "Isosceles" (two sides are congruent). Discuss the terms and place the labels below the "Triangles" card.

- Instruct kids to move their triangles to the more appropriate sub-category. Kids will move their triangles and think they're done. Check the locations and discuss any changes that might need to be made.

- Step it up. Say, "Okay, mathematicians, is there a way we can break these sub-groups into even more categories?" You may have to spill the beans yourself. Make

These triangle cards are ready to be sorted and resorted.

the labels, "Right," "Acute," and "Obtuse." Review those terms and place the cards below the "Scalene" heading.

- Explain by saying, "We're going to sort the scalene triangles into these 3 categories. If you made a scalene triangle, move it to the more specific sub-category." There's such rich discussion waiting to be held here. Discuss the attributes. Praise correct language and children who are justifying their moves.

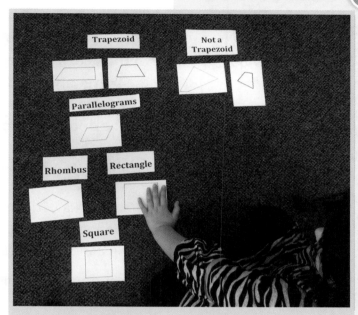

- Next place another set of index cards with the terms "Right," "Acute," and "Obtuse" under the "Isosceles" heading and ask, "Can we further categorize the triangles from this category?" Write "Equilateral" and place it under the "Acute" card. Ask, "Did anyone create an equilateral triangle?" (If not, provide the one that you prepared.) Say, "Some of the isosceles triangles can't be categorized further. They're not equilateral."

Trapezoids are now defined as quadrilaterals with *at least* 1 pair of parallel sides.

- "Okay, friends, study this arrangement we've created. Do you see how the top category, 'Triangles,' is very broad? As we move down, notice that the sub-categories become more specific and there are fewer triangles in each category. Mathematicians call this a 'hierarchy of terms.'"

- Remove all index cards and move all of the triangles back under the broad "Triangle" label. Ask, "How would things look different if in our first sort we used the sub-categories 'Acute,' 'Obtuse,' and 'Right?'" (For instance an isosceles triangle could fit under Acute, Obtuse, and Right.)

- Back in your classroom, leave the labels and triangles where students can sort and resort the shapes. Next, mount the shapes and labels on the wall or bulletin board. It makes a great visual of the attribute categories and sub-categories of triangles.

Extension: Have students sort "Quadrilaterals" (4-sided figures with 4 angles that total 360 degrees) into the sub-categories "Trapezoid" and "Not a Trapezoid" as shown in the photo.

 **QUICK TIP**

Did you know the term "trapezoid" has two different meanings? According to the older, exclusive definition, a trapezoid is a quadrilateral with <u>exactly</u> one pair of parallel sides. Using this definition, a parallelogram is not a trapezoid. The newer, inclusive definition states that a trapezoid is a quadrilateral with <u>at least</u> one pair of parallel sides. This is the definition that will most likely appear on both national CCSSM exams.

5.G.B.4 Classify two-dimensional figures in a hierarchy based on properties.

Math Practices
3 Construct Arguments & Critique Reasoning
6 Attend to Precision
7 Make Use of Structure

Follow the Clues, Sherlock!

In this activity students carefully listen to a set of clues to determine the polygon you're describing. After being inspired by your example, they write their own very specific clues about polygons (their clues will be dripping with mathematical language!), for their classmates to solve.

A pencil and a math journal are all students need for this activity.

- Say, "Okay, friends, today you're going to be like the great detective Sherlock Holmes. He put clues together and came up with amazing conclusions."

- Continue by saying, "Think back to when we categorized our triangles and quadrilaterals (refer to Sort It Out on page 214). Do you recall how each sub-category got more and more specific? Each part of my clue will be like that. The first clue will be broad, but each one will get more and more precise. I want you to listen carefully and try to picture in your mind the shape I'm describing. Ready?"

- Read these clues slowly, pausing after each one:
  Clue 1: This polygon has angles whose measurements equal 180 degrees.
  Clue 2: No two sides are congruent.
  Clue 3: The sum of 2 of the angles totals 90 degrees.

- Take a long pause. If they haven't arrived at the answer, add, "Clue 4: One angle is a right angle. What shape am I talking about?" (scalene right triangle)

- Review the clues. "The first clue tells you it's a triangle, and there are lots of triangles! The next clue lets you know it's a scalene triangle, and the third clue let's you know it's a right triangle." One example might serve the cast of characters in your room well. If there are blank stares, give more examples.

- Say, "Your task today is to write clues for other polygons. You must start with a broad definition. Each clue that follows should get more and more specific. You'll use what mathematicians call a 'hierarchy of terms.'"

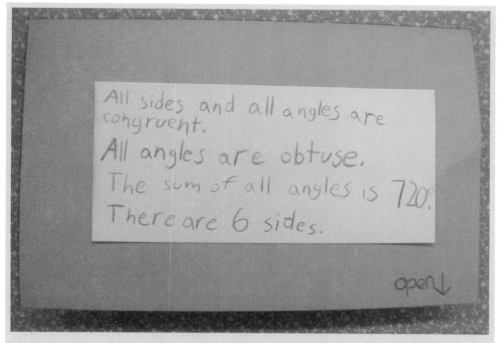

Displayed together, the student-written clues make a fun and interactive bulletin board! To check an answer, just lift the flap. (In this example, the clues describe a hexagon.)

- You may want some kids to work together. Others will be on autopilot, on their own. Encourage kids to use the vocabulary from the cards and labels that were created for Sort It Out, page 214.

- Once the clues are written in their math journals, ask kids to share them with their detective friends. This is a win-win situation for clue givers and receivers. Some eager beavers will want to write more than one clue. Rejoice!

- Your lesson may end right there or you may want students to turn edited clues into an interactive bulletin board as shown in the photo.

Copymasters & Assessments

Place-Value Chart *Use with A Milliliter Drop, page 70.*

Name: _____ Date: _____

In a multi-digit number, a digit in one place represents 10 times as much as it represents in the place to its right and $\frac{1}{10}$ of what it represents in the place to its left. (CCSSM standard 5.NBT.A.1)

Thousands	Hundreds	Tens	Ones	Tenths	Hundredths	Thousandths
Kiloliter Kilometer Kilogram	Hectoliter Hectometer Hectogram	Dekaliter Dekameter Dekagram	(base unit) **Liter** **Meter** **Gram**	Deciliter Decimeter Decigram	Centiliter Centimeter Centigram	Milliliter Millimeter Milligram
One Thousand Dollar Bill	One Hundred Dollar Bill	Ten Dollar Bill	Dollar Bill	Dime	Penny/Cent	
1,000,000 Drops	100,000 Drops	10,000 Drops	1,000 Drops	100 Drops	10 Drops	1 Drop
			A liter cube holds 1,000 drops			

Lion's Share Data Sheet *Use with The Lion's Share, page 133.*

Name: _____ Date: _____

Record what fraction of the cake each animal ate.

Elephant	Hippo	Gorilla	Tortoise	Warthog	Macaw	Frog	Ant

Write equations to represent the first 3 ways the cake is cut.

Cut #1: _____

Cut #2: _____

Cut #3: _____

Look carefully at each of the equations you recorded in the box above. What generalization can you make about what is happening to the fractions each time the cake is cut? _____

_____

_____

_____

Which Mathematical Practices did you use to solve this task?

_____

_____

Time Is Passing Data Sheet *Use with Get Up and Go!, page 140.*

Name: _____   Date: _____

 7:00 A.M.

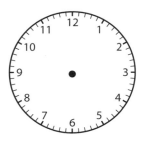 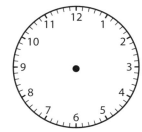 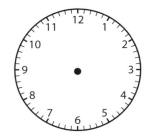 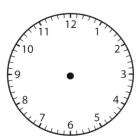

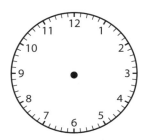 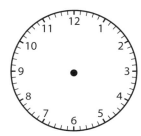 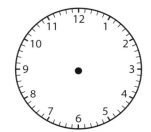 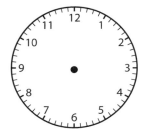

0

Copymaster • *Common Core Math in Action Grades 3–5* • © 2015 • Crystal Springs Books

Sparkles' Mix-Up Game Cards *Use with Sparkles' Mix-Up Mess!, page 166.*

Name: _____   Date: _____

Sparkles' Mix-Up Mess Card #1

OH NO! Sparkles got into the clinic and changed the heights on these children's charts. List each child's height from shortest to tallest using the same unit of measure. Show answers on a number line.

Colton: 41 inches

Dylan: 1 foot, 37 inches

Wyatt: 1 yard, 1 foot, 3 inches

Isa: $2\frac{1}{2}$ feet

Finn: $\frac{1}{2}$ yard and 36 inches

Britta: $1\frac{1}{4}$ yard

Vivian: 4 feet

Sparkles' Mix-Up Mess Card #2

OH NO! Sparkles posted these prices in the cafeteria. How confusing! List the price of each item in the cafeteria, using dollars and cents, from least to greatest. Show answers on a number line.

Milk: 4 dimes and a nickel

Chocolate Milk: 2 quarters

Cookie: 4 dimes and a penny

Hotdog: 125 cents

Fresh Fruit: 3 quarters, a dime, and 10 cents

Full Lunch: 5 quarters, 2 dimes, a nickel, and 5 pennies

Sparkles' Mix-Up Mess Card #3

OH NO! Sparkles mixed up the labels of these liquids in the science lab. Using the same unit of measure, list each liquid by amount from least to greatest. Show answers on a number line.

Lemon Juice: 1 liter and 100 milliliters

Vinegar: 2023 milliliters

Seltzer: $\frac{1}{2}$ liter and 330 milliliters

Tap Water: 6.54 liters

Rubbing Alcohol: 2450 milliliters

Red Food Dye: 430 milliliters

Yellow Food Dye: 0.02 liter

Green Food Dye: 3 milliliters

Blue Food Dye: 0.2 liter

Sparkles' Mix-Up Mess Card #4

OH NO! Sparkles changed the measurements for the Field Day Races. Using the same unit of measure, list each race in order from shortest to longest. Show answers on a number line.

Long Distance: 9,100 centimeters

3-Legged: 91 meters

Wheel Barrel: 550 centimeters

Sack Race: 460 centimeters

Crab Walk: 420 centimeters

Flag Tag Relay: 10,000 centimeters

Sneaker Relay: 43 meters

Wacky Water: 50.0 meters

Turkey Trot: 4,800 centimeters

Grab Bag: 5,200 centimeters

Make a Peacock *Use with Peacock Power!, page 206.*

Name: _____ Date: _____

Read all directions carefully. Your score will be based on correct measurements and shapes.

	Measurements and Directions	**Points**
Body	An isosceles triangle with a base of 2 inches and a height of 6 inches.	**/10**
6 Tail Feathers	Feathers should be different colors. Construct and cut 6 right triangles: • 2 with 2-inch bases and a height of 5 inches. • 2 with $1\frac{1}{2}$-inch bases and a height of 4 inches. • 2 with 1-inch bases and a height of 3 inches. Arrange the feathers symmetrically so there are 3 feathers on each side of the body.	**/60**
Head	A circle with a 2-inch diameter. Circle eyes.	**/10**
Feet	Two small congruent right triangles.	**/10**
Beak	Cut a rhombus. Fold it in half along the obtuse angles, forming 2 triangles. Glue down 1 triangle so mouth appears open.	**/5**
Crest	A triangle in which no side is more than $1\frac{1}{2}$ inches. Fringe the top side; glue to the top of the head.	**/5**
Total		**/100**

 Copymaster • *Common Core Math in Action Grades 3–5* • © 2015 • Crystal Springs Books

See What They Know: Grade 3 Assessments

We all know that formative assessment is essential to good teaching. As you work with the standards in each domain, consider having your students respond to the prompts below to see just how much their skills have grown. To track progress, you may want to copy the prompts, cut them apart, have students paste each strip in their math journals and respond there.

Make Sevens Easy

Use the distributive property to solve 8×7. Indicate the decomposed factors, or use an array to show your thinking.

Assessment OA-3

A Trip to the Park

Theme park tickets cost $37 for children. Britt's mom needs to buy 6 tickets for her children and their friends. Use rounding to compute about how much money it will cost.

Assessment NBT-3

Pizza by the Slice

Adam has $\frac{1}{3}$ of a pizza, and his sister Sarah has $\frac{2}{6}$ of that pizza. Sarah thinks she has the most because she has sixths and 6 is a BIG number. Adam keeps telling her that she does not have more pizza. Who is right and why? Use fraction models and a number line to show your reasoning. Your answer should include labels and precise language.

Assessment F-3

Figure This

Roanne made the rectilinear figure pictured here. She left some sides without labels. Figure out and fill in the inch measurements that belong in each box. What are the perimeter and area of this shape?

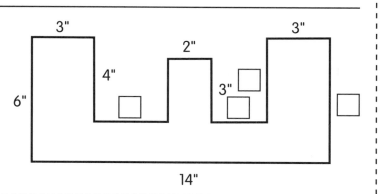

Assessment MD-3

✂ -

Compare to Show Understanding

How is a square like a rhombus? (Give at least 2 ways.)

✂ -

Grade 3 Assessment Answer Key

Examples of Possible Solutions:

Make Sevens Easy: In order to solve this problem using the distributive property, students must decompose or break apart one of its factors. For example, a student could decompose the 8 into 5 and 3 and explain: "5 × 7 = 35 and 3 × 7 = 21, and 35 + 21 = 56."

A Trip to the Park: A student could round the cost of each ticket to $40 (rounding to $30 will arrive at too low an estimate). Next, since 6 tickets are needed, the student could figure: "6 × $4 = $24," and then reason: "6 × $40 = $240." While $240 is more than the amount needed, the problem asks "about" how much tickets will cost.

Pizza by the Slice: Adam is correct. A student could draw circles or rectangles divided into thirds and sixths and shade in $\frac{1}{3}$ and $\frac{2}{6}$ to show equal amounts of pizza. A student could also draw a number line showing that the distance from 0 to $\frac{2}{6}$ is equivalent to the distance from 0 to $\frac{1}{3}$. Look for labels and precise language.

Figure This: Starting at the bottom left side of the figure and going clockwise, the missing numbers are: 3, 3, 4, and 6. Perimeter is 54 inches (the sum of all of the sides). Area can be figured by looking at this shape as a series of 4 rectangles. The long rectangle at the bottom is 14 × 2 inches. The 2 rectangles on the right and left are each 3 × 4 inches. The middle rectangle is 2 × 3 inches. Area: (14 × 2) + (3 × 4) + (3 × 4) + (2 × 3) = 58 square inches.

Compare to Show Understanding: A square is a type of rhombus. A rhombus has 4 congruent sides and its opposite sides are parallel. All squares are rhombi, but rhombi that don't have right angles aren't squares. Insist on more than 1 definition, and look for precise language. Children may use drawings with labels. Praise correct labels and illustrations that were drawn using tools such as straightedges or rulers.

See What They Know: Grade 4 Assessments

We all know that formative assessment is essential to good teaching. As you work with the standards in each domain, consider having your students respond to the prompts below to see just how much their skills have grown. To track progress, you may want to copy the prompts, cut them apart, have students paste each strip in their math journals and respond there.

Compare & Take an Extra Step

Amanda weighs 93 pounds. She weighs 3 times as much as her little sister. What is the combined weight of the 2 girls?

Assessment OA-4

Show It Two Ways

Calculate 4×112, and draw a picture of this problem using base-10 blocks or an area model.

Assessment NBT-4

Lemonade Stand

Lucy sold $\frac{1}{2}$ of the lemonade she made at her lemonade stand, and Larry sold $\frac{5}{6}$ of the lemonade he made at his stand. Who sold more? Use pictures, words, and symbols to explain your answer.

Assessment F-4

Stepping onto Patterns

Susan is looking at a beautiful tile floor made with 3 different regular polygons. In this floor she sees regular hexagons with squares on each of the 6 sides of the hexagon. What regular polygon would have to be in the space between each square?

You may use pattern blocks to model this tile floor, and then draw the floor. Choose any point on your drawing and label the degrees in the angles. What is the total around that point?

Assessment MD-4

Assessment G-4

Attend to Precision

Draw a polygon. Write the name for your 2-D shape. Label each angle as right, acute, or obtuse. Label the line segments and note if your shape has parallel or perpendicular lines.

Grade 4 Assessment Answer Key

Examples of Possible Solutions:

Compare & Take an Extra Step: A student might solve this problem using number model sentences. For example, "93 ÷ 3 = 31" determines the little sister's weight, and "93 + 31 = 124" determines the combined weight of the 2 girls. If a student chooses model drawing, then Amanda's unit bar will be 3 times the length of the little sister's unit bar.

Show It Two Ways: 4 × 112 = 448. A correct student drawing using base-10 blocks will show 4 flats, 4 longs, and 8 ones.

Lemonade Stand: Students must recognize that there isn't enough information to solve this problem. They would need to know the amount of lemonade each child had at the start.

Stepping onto Patterns: Triangles are the shapes between each square. Around any given point, a student should label the angle in the hexagon 120°, the angle in the triangle 60°, and 2 of the angles in the square 90° each. 120° + 60° + 90° + 90° = 360°.

Attend to Precision: Shapes will vary, but all angles, line segments, and lines *must* be labeled.

We all know that formative assessment is essential to good teaching. As you work with the standards in each domain, consider having your students respond to the prompts below to see just how much their skills have grown. To track progress, you may want to copy the prompts, cut them apart, have students paste each strip in their math journals and respond there.

Express Yourself

Evaluate this expression: $2 \times [5 + (3 \times 4)]$.

Area for Understanding

Find the quotient using a rectangular array or area model: $299 \div 13 = ?$

The Pizza Problem

Marcia ate the following fractions of a medium pizza each day for lunch this week:

Monday: $\frac{1}{3}$ Tuesday: $\frac{1}{2}$ Wednesday: $\frac{1}{4}$ Thursday: $\frac{2}{6}$ Friday: $\frac{2}{3}$

Her brother John ate the rest each day.

1. Which 2 days did Marcia eat the equivalent amount? Prove your answer with visual models.

2. Rank the amounts she ate from least to greatest using a visual model. Then place the fractions on a number line.

3. Did Marcia eat more or less pizza than John? How do you know without calculating? Draw/illustrate your thinking, explain your thinking in words, and write the math expression to show your answer.

Assessment OA-5

Assessment NBT-5

Assessment F-5

See What They Know: Grade 5 Assessments
We all know that formative assessment is essential to good teaching. As you work with the standards in each domain, consider having your students respond to the prompts below to see just how much their skills have grown. To track progress, you may want to copy the prompts, cut them apart, have students paste each strip in their math journals and respond there.

Assessment MD-5

Pentomino Volume

The 3-D pentomino pictured to the right is made up of right rectangular prisms. If one edge measures 2 centimeters, what is the volume of this solid?

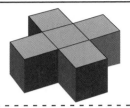

Assessment G-5

What Is a Square?

Sarah says that a square is a rhombus, but Judd disagrees. He says a rhombus can't have all equal angles. Who is correct and why? Use words and a diagram to show your thinking. Your answer should include labels and precise language.

Grade 5 Assessment Answer Key

Examples of Possible Solutions:

Express Yourself: Students must understand the order of operations in order to solve this problem. Step 1: "3 × 4 = 12." Step 2: "5 + 12 = 17." Step 3: "17 × 2 = 34."

Area for Understanding: 299 ÷ 13 = 23. Students should solve with an area model or array.

The Pizza Problem: (1) Monday and Thursday. (2) $\frac{1}{4}, \frac{1}{3}$ and $\frac{2}{6}, \frac{1}{2}, \frac{2}{3}$. (3) Marcia ate less. Here's one way a student might explain: "On Tuesday Marcia ate the same amount of pizza as John. On 3 other days she ate much less pizza than John. On Friday she ate just $\frac{1}{3}$ more pizza than John. It's easy to see that Marcia ate less pizza than her brother." Math equations: Marcia $\left(\frac{1}{3} + \frac{1}{2} + \frac{1}{4} + \frac{2}{6} + \frac{2}{3} = 2\frac{1}{12}\right)$ and John $\left(\frac{2}{3} + \frac{1}{2} + \frac{3}{4} + \frac{4}{6} + \frac{1}{3} = 2\frac{11}{12}\right)$.

Pentomino Volume: 2 × 2 × 2 = 8 cubic cm for each cube. Five of these cubes are equal to 40 cubic centimeters.

What Is a Square?: Sarah is correct. All squares are rhombi, but not all rhombi are squares. Student explanations and/or diagrams should show that rhombi have 4 congruent sides and opposite sides are parallel. Diagrams of squares should show right angles and 4 congruent sides. Labels of sides and correct terms are expected in diagrams.

References & Resources

Professional Books/Resources

Chambers, Donald L., ed. 2002. *Putting Research into Practice in the Elementary Grades: Readings from Journals of the NCTM*. Reston, VA: National Council of Teachers of Mathematics.

Gomez, Emiliano. 2009. *Why Are Fractions So Useful at Predicting Success in Math?* Talk presented at the California Mathematics Council Annual Conference, Monterey, CA.

Hiebert, James, et al. 2000. Making Sense: *Teaching and Learning Mathematics with Understanding*. Portsmouth, NH: Heinemann.

Leinwand, Steven. 2009. *Accessible Mathematics: 10 Instructional Shifts that Raise Student Achievement*. Portsmouth, NH: Heinemann.

Marzano, Robert J. 2004. *Building Background Knowledge for Academic Achievement: Research on What Works in Schools*. Alexandria, VA: ASCD.

McNamara, Julie, and Meghan M. Shaughnessy. 2010. *Beyond Pizzas & Pies: 10 Essential Strategies for Supporting Fraction Sense*. Sausalito, CA: Math Solutions.

Small, Marian. 2012. *Good Questions: Great Ways to Differentiate Mathematics Instruction*. 2nd ed. New York: Teachers College Press.

———. 2013. *Eyes on Math: A Visual Approach to Teaching Math Concepts*. New York: Teachers College Press.

Smith, Margaret. 2013. "How Can We Use High-Level Tasks to Promote Equity in the Mathematics Classroom?" (website) ifl.lrdc.pitt.edu/index.php/resources/ask_the_educator/peg_smith.

The Common Core Standards Writing Team. 2012. "Progression for the Common Core State Standards Mathematics, Geometry K–6 (draft)." (website) commoncoretools.files.wordpress.com/2012/06/ccss_progression_g_k6_2012_06_27.pdf.

Van de Walle, John A., Karen S. Karp, and Jennifer Bay-Williams. 2003. *Elementary and Middle School Mathematics: Teaching Developmentally*. 8th ed. Upper Saddle River, NJ: Pearson.

Whitin, Phyllis, and David J. Whitin. 2000. *Math Is a Language Too: Talking and Writing in the Mathematics Classroom*. Urbana, IL: National Council of Teachers of English.

Yee, Lee Peng, and Lee Ngan Hoe. 2009. *Teaching Primary School Mathematics: A Resource Book*. 2nd ed. Singapore: McGraw Hill Education (Asia).

Children's Literature with Math Connections for Grades 3–5

The following children's books, appropriate for grades 3–5, are referenced within activities in this book.

Operations and Algebraic Thinking

Amanda Bean's Amazing Dream by Cindy Neuschwander
Anno's Magic Seeds by Mitsumasa Anno
Centipede's 100 Shoes by Tony Ross
Cheetah Math: Learning About Division from Baby Cheetahs by Ann Whitehead Nagda
Delicious: The Art and Life of Wayne Thiebaud by Susan Goldman Rubin
Martha Blah Blah by Susan Meddaugh
One Is a Snail, Ten Is a Crab by April Pulley Sayre
Paul Klee (Getting to Know the World's Greatest Artists) by Mike Venezia
Six-Dinner Sid by Inga Moore
Two Ways to Count to Ten, a Liberian Folk Tale by Ruby Dee

Number and Operations in Base Ten

A Remainder of One by Elinor J. Pinczes
Equal Shmequal by Virginia Kroll
If America Were a Village by David J. Smith

Number and Operations—Fractions

Fraction Fun by David A. Adler
Full House: An Invitation to Fractions by Dayle Ann Dodds
Multiplying Menace, the Revenge of Rumpelstiltskin by Pam Calvert
Paul Klee (Getting to Know the World's Greatest Artists) by Mike Venezia
Picture Pie by Ed Emberly
The Doorbell Rang by Pat Hutchens
The Lion's Share: A Tale of Halving Cake and Eating It, Too by Matthew McElligott

Measurement and Data

Bigger, Better, Best! by Stuart J. Murphy
Get Up and Go! by Stuart J. Murphy
Hamster Champs by Stuart J. Murphy
Jack and the Beanstalk, any version
Mummy Math: An Adventure in Geometry by Cindy Neuschwander

Pablo Picasso: Breaking All the Rules (Smart About Art) by True Kelley
Pastry School in Paris: an Adventure in Capacity by Cindy Neuschwander
Smithsonian Handbooks: Butterflies and Moths by David Carter
The Wright 3 by Blue Balliett
Vincent Van Gogh: Sunflowers and Swirly Stars (Smart About Art) by Joan Holub

Geometry

A Fly on the Ceiling by Julie Glass
Coyote Places the Stars by Harriet Peck Taylor
If You Were a Quadrilateral by Molly Blaisdell
Sweet Clara and the Freedom Quilt by Deborah Hopkinson
The Greedy Triangle by Marilyn Burns
The Keeping Quilt by Patricia Polacco

Websites

Catherine Kuhns
www.catherinekuhns.com

Common Core Conversation
www.commoncoreconversation.com/math-resources

Common Core State Standards Initiative
www.corestandards.org

Dan Meyer
blog.mrmeyer.com

Math Is Fun
www.mathisfun.com

The Mathematics Common Core Toolbox
www.ccsstoolbox.com

MC2, Marrie Lasater & Cindy Cliche
www.mathmc2.com

National Gallery of Art
images.nga.gov

NCTM Illuminations/Dynamic paper
illuminations.nctm.org (use site's search function to find Dynamic Paper)

Shodor
www.shodor.org

Index

Underscored page numbers indicate boxed text.
Italicized page numbers indicate copymasters.

A

Abstract understanding, 3, 7–8, <u>8</u>, 15
Acute angles, 139, 172, 173, 178, 204, 215
Add Comparison to Your Skills, 32–33
Adding-machine tape, for fractional number line, 97
Addition
    grade 3
        in measurement and data domain, 148
        in operations and algebraic thinking domain, 29
    grade 4
        in measurement and data domain, 180, 181
        in number and operations—fractions domain, 98,
            99–100, 104, 105–6, 113, 115–16
        in number and operations in base ten domain, 61, 62
    grade 5, in number and operations—fractions domain,
        117–19, 119–20
Addition grid, for seeing patterns, 30
Algebra. *See also* Operations and algebraic thinking
    fractions as preparation for, 81
Amanda Bean's Amazing Dream, 25–26, 34–35
Amazing Arrays, 17
Anchor charts, as memory aid, <u>123</u>
Angle rulers, 9, 174, 176, 178
Angles
    acute, 139, 172, 173, 178, 204, 215
    in geometry domain, 199
    measuring, 172–81
    obtuse, 139, 173, 204, 215
Angles in Plates, 174–75
Angles of a Crown, 172–73
Anno's Magic Seeds, 36
Another Serving of Cake, Please, 98–99
Answer keys, for assessments. *See* Assessments
Answers, taking time for, 11
Ant's Progress, The, 182–84
Area, 139
    with fractional numbers, 124–26
    measurement of, 148–59, 160, 161, 162–63, 168–69,
        196
    perimeter and, 159, 160, 161, 162–63
Area for Understanding, for Grade 5 assessment, *229, 230*
Area model
    for division problem, 66
    for multiplication, 63–64
Area Model Method—Multiplication, 63–64
Arrays
    in division, 20
    in multiplication, 16, 17, 19, 24, 85
Arrays, Amazing, 17

Artwork
    arrays in, 16–17, 85–86
    measuring angles in, 176–77
Assessments
    Grade 3, *225–26*
        answer key, 226
    Grade 4, *227–28*
        answer key, 228
    Grade 5, *229–30*
        answer key, 230
    purpose of, 5
    using, 5
Associative property, 23–24, 192, 193
Attend to Precision, for Grade 4 assessment, 228, *228*

B

Bar graph, 144–46
Base-10 blocks
    for division problems, 28, 65, 66
    for measuring distance, 182, 183, 184
    for modeling multiplication, 54
    for teaching decimals, 74, 114
Base ten. *See* Number and operations in base ten
Beans, Handfuls of, 144–46
Bigger, Better, Best (activity), 148–49
Bigger, Better, Best! (book), 148–49, 150–51
Blueprints, 205
Box Turtles, Building Fractions with, 117–19
Braces. *See* Brackets
Brackets, 44, 45
Bring Out the Coins, 67–68
Bruner, Jerome, <u>8</u>
Build a Liter, 142
Building Cubes, 186–87
Building Fractions with Box Turtles, 117–19

C

Cake, Another Serving of, Please, 98–99
Cake, It's a Piece of, 95–96
CCSS for Mathematics, 3
CCSSI, 6
CCSSM. *See* Common Core State Standards in Mathematics
Centimeters, 74–75, 164, 165, 182, 183, 184, 188, 189
Centipede's 100 Shoes (activity), 21–22
Centipede's 100 Shoes (book), 21–22
Cheetah Math, 27–28
Cheetah Math: Learning About Division from Baby Cheetahs,
    27–28
Children's literature, activities using. *See* Literature, children's,
    activities using
Circles
    fractions as part of, 81
    in geometry domain, 207

Cluster, activities organized by, 3
Cluster codes, CCSSM, 3
Coins, Bring Out the, 67–68
Common Core, 8, 10, 11, 13
Common Core State Standards Initiative (CCSSI), 6
Common Core State Standards in Mathematics (CCSSM)
    activities supporting, 6
    cluster codes, 3
    core skills of, 49
    domains in, 2
    history of, 6
    identified for activities, 3
    tests, using correct terminology in, 13, 189, 215
Commutative property, 15, 23, 128
Compare & Take an Extra Step, for Grade 4 assessment, *227*, 228
Compare to Show Understanding, for Grade 3 assessment, 226, *226*
Comparisons, in grade 4 activity, 32–33
Compass, 9, 207
Compensation Strategy, The, 52–53
Concrete, pictorial, abstract (C–P–A) reasoning, 3, 7–8, 8, 15
Constellations, plotting, on coordinate grid, 212–13
Construction Site, 204–5
Content standards, activities organized by, 2, 3
Coordinate grids, 47–48, 199, 210–13
Copymasters, 5
    Lion's Share Data Sheet, 133, 134, *221*
    Make a Peacock, 206, 207, *224*
    Place-Value Chart, 70, *220*
    Sparkles' Mix-Up Game Cards, 166, 167, *223*
    Time Is Passing Data Sheet, 140, 141, *222*
Counting Cubes, 188–90
Counting Feet, 42–43
Cover It!, 150–51
Coyote Places the Stars, 212–13
Coyote's Night Sky, 212–13
CPA icon, 8
C-P-A stages of reasoning, 3, 7–8, 8, 15
Crown, Angles of a, 172–73
Cubed and Then Counted, 190–91
Cubes, 9
    for modeling division, 27
    for understanding volume, 186–98
Cubes, Building, 186–87
Cubes, Counting, 188–90
Cubes, Hoops & Smiles!, 192–93

D

Data. *See also* Measurement and data
    on bar graph, 144–46
    on line plots, 144–47, 169–71, 184–85
Decimals
    adding, 115–16
    common misconception about, 73
    comparing, 72–73

    expressing fractions as, 114–16
    in menu prices, 79–80
    multiplied by whole number, 68–69
    reading and writing to thousandths, 70–71
    rounding, 74–75
Decimeters, 74, 182, 183, 184
Decomposing Numbers, 44–45
Decomposing numbers
    in area problem, 156–57
    expanded form for, 57
    in fractions, 101, 102–3
    for modeling distributive property, 24
    for multiplication, 63, 64
    parentheses and brackets for, 44–45
    with pattern blocks, 122–23
Decomposing rectilinear shapes, 148
Delicious: The Art and Life of Wayne Thiebaud, 16
Denominators
    grade 3 understanding of
        activities for, 82, 83–85, 94
        expectations for, 81
    grade 4 understanding of
        activities for, 99–100, 112, 113
        expectations for, 81
    grade 5 understanding of
        activities for, 119–20, 129–30
        expectations for, 81
Descartes, René, 210
Dessert Time, Old-Fashioned, 137–38
Detectives, Measurement, 159–61
Dienes, Zoltan, 8
Distance
    estimating, 164–65
    measuring, 182–84
Distribute and Conquer, 156–57
Distributive property, 24, 156, 157
Division
    arrays in, 20
    grade 3, in operations and algebraic thinking domain, 18–19, 20, 21–22, 25–26, 27–28, 29
    grade 4
        in number and operations in base ten domain, 65–66
        in operations and algebraic thinking domain, 33
    grade 5
        in number and operations—fractions domain, 120–21, 133–34, 135–36, 137–38
        in number and operations in base ten domain, 77–78
Division, Dreams of, 25–26
Divvy Up!, 18–19
Does Perimeter Determine Area?, 162–63
Domain sections. *See also* Geometry; Measurement and data; Number and Operations—Fractions; Number and Operations in Base Ten; Operations and algebraic thinking
    color-coded grades in, 3
    organization of, in this book, 2

Doorbell Rang (activity), 120–21
Doorbell Rang, The (book), 120–21
Doubling strategy, for pattern practice, 30, 31
Drawing pictures, in pictorial stage of reasoning, 7
Dreams of Division, 25–26
Dynamic Paper tool, 9

E

Educators and educational experts
    Bruner, Jerome, 8
    Dienes, Zoltan, 8
    Meyer, Dan, 78
    Montessori, Maria, 8
    Piaget, Jean, 8
    Renzulli, Joseph, 12
    Small, Marian, 10
"Equal," 58
Equal Shmequal (activity), 57–58
Equal Shmequal (book), 57–58
Equivalent fractions
    grade 4 activities for understanding, 95–96, 97, 112
    grade 5 activities for understanding, 117–19
Estimating, 36, 59, 144–46, 164–65
Expanded form
    decomposing numbers into, 57, 58
    demonstrating, 71
Express Yourself, 45–46
    for Grade 5 assessment, *229*, 230
Extensions, purpose of, 4

F

Factors, in Grade 4 activity, 37–39
Fairy tales, 154–55
Famous People & Places, 154–55
Feet, Counting, 42–43
Figure This, for Grade 3 assessment, *225*, 226
Fishy Bowls, 112
Fly on the Ceiling, A, 210–11
Follow the Clues, Sherlock!, 216–17
Formative assessment. *See* Assessments
Fraction Fun (activity), 93–94
Fraction Fun (book), 93–94
Fraction Rulers, 102–3, 104, 105, 106, 106, 137, 138
Fractions. *See also* Number and operations—fractions
    angle ruler for understanding, 174–75
    describing shapes, 202, 203
    on line plots, 146–47, 169, 170, 171, 184–85
Fractions, Jump Rope, 87–88
Fractions, Mr. Franklin's, 135–36
Fractions, Paper-Folded, 83–85
Fractions, Paper Plate, 88–90
Fractions, Reasoning with, 129–30
Fraction Stories & Rulers, 105–7
Fraction Teams, 97
Full House, 110–11

Full House: An Invitation to Fractions, 110–11

G

Garden, creating visual model of, 115–16
Geometric shape patterns, building, 40–41
Geometric solids, 9
Geometry, 199–217
    Grade 3, 199, 200–203
        Paper Quilt, 202–3
        Quadrilaterals, 200–201
    Grade 4, 199, 204–9
        Construction Site, 204–5
        Hawaiian Quilt, 208–9
        Peacock Power!, 206–8, *224*
    Grade 5, 199, 210–17
        Coyote's Night Sky, 212–13
        Follow the Clues, Sherlock!, 216–17
        Move It!, 210–11
        Sort It Out, 214–15
Getting to Know the World's Greatest Artists, 177
Get Up and Go! (activity), 140–41, *222*
Get Up and Go! (book), 140–41
Go Figure, Father Time, 77–78
Golf, Hamster, 164–65
Goniometer, 174
Grade 3 activities by domain
    geometry, 199, 200–203
        Paper Quilt, 202–3
        Quadrilaterals, 200–201
    measurement and data, 140–63
        Bigger, Better, Best, 148–49
        Build a Liter, 142
        Cover It!, 150–51
        Distribute and Conquer, 156–57
        Does Perimeter Determine Area?, 162–63
        Famous People & Places, 154–55
        Get Up and Go!, 140–41, *222*
        Handfuls of Beans, 144–46
        Hand It to Me!, 146–47
        Liter Benchmarks, 143
        Measurement Detectives, 159–61
        A Rectilinear Place to Play, 158–59
        There's a Formula for That!, 152–53
    number and operations—fractions, 81, 82–94
        Fraction Fun, 93–94
        Jump Rope Fractions, 87–88
        Mr. Klee's Squares, 85–86
        My Friends & Me, 82–83
        My Whole Hexagon, 90–91
        Paper-Folded Fractions, 83–85
        Paper Plate Fractions, 88–90
        A Ribbon Ruler Tool, 86–87
        Stacking Wholes, 92–93
    number and operations in base ten, 50–54
        The Compensation Strategy, 52–53
        Hundred Charts, 54

Grade 3 activities by domain (continued)
    Race Car Rounding, 50–51
  operations and algebraic thinking, 15, 16–31
    Amazing Arrays, 17
    Centipede's 100 Shoes, 21–22
    Cheetah Math, 27–28
    Divvy Up!, 18–19
    Dreams of Division, 25–26
    Martha Sells Letters for $100, 28–29
    Mix It Up!, 23–24
    Multiplication Masterpieces, 16–17
    Practicing with Patterns, 30–31
    Six-Dinner Sid, 19–20
Grade 3 assessments, *225–26*
  answer key, 226
Grade 4 activities by domain
  geometry, 199, 204–9
    Construction Site, 204–5
    Hawaiian Quilt, 208–9
    Peacock Power!, 206–8, *224*
  measurement and data, 164–81
    Angles in Plates, 174–75
    Angles of a Crown, 172–73
    Hamster Champs, 180–81
    Hamster Golf, 164–65
    Measuring Masterpieces, 176–77
    Protractor Measurement, 178–79
    Sparkles' Mix-Up Mess!, 166–67, *223*
    Storybook Character Problems, 168–69
    Wing It!, 169–71
  number and operations—fractions, 81, 95–116
    Another Serving of Cake, Please, 98–99
    Fishy Bowls, 112
    Fraction Rulers, 102–3
    Fraction Stories & Rulers, 105–7
    Fraction Teams, 97
    Full House, 110–11
    How Does Your Garden Grow?, 115–16
    It's a Piece of Cake, 95–96
    Let the Games Begin!, 104–5
    Like Denominator Addition, 99–100
    Loops & Loops, 113
    Multiply with Pattern Blocks, 107–8
    Pattern Blocks, Again!, 108–9
    Powerful Kid-Created Problems, 111
    Roll to a Meter, 114–15
    What's the Whole?, 101
  number and operations in base ten, 55–66
    Area Model Method—Multiplication,
      63–64
    Equal Shmequal, 57–58
    Place-Value Rounding, 59–60
    Puzzle It Out, 61–62
    A Remainder of One, 65–66
    Ten Times the Money, 55–56
    Think & Fill in the Blank, 62

  operations and algebraic thinking, 15, 32–41
    Add Comparison to Your Skills, 32–33
    More Than Two Ways, 37–39
    Multiply—How & Why, 34–35
    Problems with Seeds, 36
    Toothpick Triangles, 40–41
Grade 4 assessments, *227–28*
  answer key, 228
Grade 5 activities by domain
  geometry, 199, 210–17
    Coyote's Night Sky, 212–13
    Follow the Clues, Sherlock!, 216–17
    Move It!, 210–11
    Sort It Out, 214–15
  measurement and data, 182–98
    The Ant's Progress, 182–84
    Building Cubes, 186–87
    Counting Cubes, 188–90
    Cubed and Then Counted, 190–91
    Cubes, Hoops & Smiles!, 192–93
    Lemon Juice Line-Up, 184–85
    Mummies and Math, 194–96
    Pentominoes, 197–98
  number and operations—fractions, 81, 117–38
    Building Fractions with Box Turtles, 117–19
    Doorbell Rang, 120–21
    The Lion's Share, 133–34, *221*
    Mr. Franklin's Fractions, 135–36
    Old-Fashioned Dessert Time, 137–38
    Painting Turtles, 131–32
    Paper Plates Show the Way, 127–28
    Reasoning with Fractions, 129–30
    Thinking About Groups, 122–23
    Tiles to the Rescue, 124–26
    Unlike Denominator Addition, 119–20
  number and operations in base ten, 67–80
    Bring Out the Coins, 67–68
    Go Figure, Father Time, 77–78
    If America Were a Village, 72–73
    May I Take Your Order?, 79–80
    A Milliliter Drop, 70–71, *220*
    No Problem!, 75–77
    To the Point, 68–69
    Rounding Decimals, 74–75
  operations and algebraic thinking, 15, 42–48
    Brackets, 44
    Counting Feet, 42–43
    Decomposing Numbers, 44–45
    Express Yourself, 45–46
    Patterns with Rules, 47–48, <u>213</u>
Grade 5 assessments, *229–30*
  answer key, 230
Grade level. *See also grade levels 3 through 5*
  activities organized by, 2
Graphs, 139
  bar, 144–46

Graphs *(continued)*
    line plots, 139, 144–47, 184–85
"Greater than," 58
Greater than symbol, 73
Greedy Triangle, The, 204–5
Groups, Thinking About, 122–23
Group size suggestions, for activities, 3

H

Hamster Champs (activity), 180–81
Hamster Champs (book), 180–81
Hamster Golf, 164–65
Handfuls of Beans, 144–46
Hand It to Me!, 146–47
Hawaiian Quilt, 208–9
Hexagon, My Whole, 90–91
Hexominoes, 198
Hierarchy of terms, 216
How Does Your Garden Grow?, 115–16
How to use this book, 2–5
Hundred chart
    for seeing patterns, 30
    skip counting on, 54
Hundred Charts, 54
Hypotenuse, 207

I

If America Were a Village (activity), 72–73
If America Were a Village (book), 72–73
If You Were a Quadrilateral, 200–201
Input-output charts, for pattern practice, 30, 31
It's a Piece of Cake, 95–96

J

Jack and the Beanstalk, 144–45
Jump rope, as number line, 87–88, 89
Jump Rope Fractions, 87–88

K

Keeping Quilt, The, 202–3

L

Language. *See also* Vocabulary terms
    interchangeable, in math conversations, 13
Lemonade Stand, for Grade 4 assessment, *227,* 228
Lemon Juice Line-Up, 184–85
Length, measuring, 146–47
"Less than," 58
Let the Games Begin!, 104–5
Like Denominator Addition, 99–100, 118
Line, in geometry domain, 204
Line plots, 139, 144–47, 169–71, 184–85
Line segments, 204, 212, 213
Lion's Share, The, 133–34, *221*
Lion's Share, The: A Tale of Halving Cake and Eating It, Too,
    133–34

Lion's Share Data Sheet, 133, 134, *221*
Liquid measurement, 70–71, 139, 142, 143
Liter, 70–71, 142, 143, 184–85
Liter, Build a, 142
Literature, children's, activities using, 4
    Bigger, Better, Best, 148–49
    Building Fractions with Box Turtles, 117–19
    Centipede's 100 Shoes, 21–22
    Cheetah Math, 27–28
    Construction Site, 204–5
    Counting Feet, 42–43
    Cover It!, 150–51
    Coyote's Night Sky, 212–13
    Doorbell Rang, 120–21
    Dreams of Division, 25–26
    Equal Shmequal, 57–58
    Fraction Fun, 93–94
    Full House, 110–11
    Get Up and Go!, 140–41
    Hamster Champs, 180–81
    Handfuls of Beans, 144–45
    If America Were a Village, 72–73
    The Lion's Share, 133–34
    Martha Sells Letters for $100, 28–29
    Measuring Masterpieces, 176–77
    More Than Two Ways, 37–39
    Move It!, 210–11
    Mr. Klee's Squares, 85–86
    Multiplication Masterpieces, 16–17
    Multiply—How & Why, 34–35
    Mummies and Math, 194–96
    Paper Quilt, 202–3
    Pattern Blocks, Again!, 109
    Pentominoes, 197–98
    Problems with Seeds, 36
    Quadrilaterals, 200–201
    A Remainder of One, 65–66
    Six-Dinner Sid, 19–20
    Wing It!, 169–71
Liter Benchmarks, 143
Loops & Loops, 113

M

Make a Peacock, 206, 207, *224*
Make Sevens Easy, for Grade 3 assessment, *225,* 226
Manipulatives, in concrete stage of reasoning, 7
Martha Blah Blah, 28–29
Martha Sells Letters for $100, 28–29
Masterpieces, Measuring, 176–77
Masterpieces, Multiplication, 16–17
Mathematical practice standards. *See* Standards for
    Mathematical Practice
Mathematical writing, 12, 13. *See also* Math journals
Mathematics instruction, keys to meaningful, 11–12
Math games
    for adding and subtracting mixed numbers, 104–5

Math games (continued)
 online, <u>66</u>
Math journals
 assessment strips in, 225, 226, 227, 228, 229, 230
 benefits of, 4
 purpose of, 13
 tools needed in, 30
Math Meetings, 5, 78
Math talk, 12, <u>13</u>. *See also* Vocabulary terms
May I Take Your Order?, 79–80
Measurement
 of angles, 172–81
 area, 124–26, 148–59, 160, 161, 162–63, 168–69, 196
 converting larger units to smaller units, 164–65
 converting like measurement units, 182–84
 distance, 164–65, 182–84
 length, 146–47
 linear, 142
 liquid, 70–71, 139, 142, 143
 metric (*see specific metric units of measurement*)
 perimeter, 159–61, 162–63, 168–69, 195–96, 197–98
 time, 77–78, 140–41
 of unlike units, 166–67
 volume, 142, 143, 186–98
Measurement and data, 139–98
 Grade 3, 140–63
  Bigger, Better, Best, 148–49
  Build a Liter, 142
  Cover It!, 150–51
  Distribute and Conquer, 156–57
  Does Perimeter Determine Area?, 162–63
  Famous People & Places, 154–55
  Get Up and Go!, 140–41, *222*
  Handfuls of Beans, 144–46
  Hand It to Me!, 146–47
  Liter Benchmarks, 143
  Measurement Detectives, 159–61
  A Rectilinear Place to Play, 158–59
  There's a Formula for That!, 152–53
 Grade 4, 164–81
  Angles in Plates, 174–75
  Angles of a Crown, 172–73
  Hamster Champs, 180–81
  Hamster Golf, 164–65
  Measuring Masterpieces, 176–77
  Protractor Measurement, 178–79
  Sparkles' Mix-Up Mess!, 166–67, *223*
  Storybook Character Problems, 168–69
  Wing It!, 169–71
 Grade 5, 182–98
  The Ant's Progress, 182–84
  Building Cubes, 186–87
  Counting Cubes, 188–90
  Cubed and Then Counted, 190–91
  Cubes, Hoops & Smiles!, 192–93
  Lemon Juice Line-Up, 184–85

  Mummies and Math, 194–96
  Pentominoes, 197–98
Measurement Detectives, 159–61
Measuring Masterpieces, 176–77
Measuring tapes, 9
Mental math
 for grade 3 activities, 28, 29, 140
 for grade 4 activities, 32, 165
 warm-ups for, 71
Menus, for decimals practice, 79–80
Meter, Roll to a, 114–15
Meters, 165, 182, 183, 184
Meter stick
 for measuring distance, 182, 183, 184
 for teaching decimals, 74, 75, 114–15
Metric measurements. *See* specific metric units of measurement
Meyer, Dan, <u>78</u>
Milliliter Drop, A, 70–71, *220*
Millimeters, 74, 75, 182, 183
Mix It Up!, 23–24
Money, Ten Times the, 55–56
Money activities, for understanding place value, 55–56, 67–68,
 79–80
Montessori, Maria, <u>8</u>
Moon Cake recipe, 138
More Than Two Ways, 37–39
Move It!, 210–11
Mr. Franklin's Fractions, 135–36
Mr. Klee's Squares, 85–86
Multi-digit whole numbers
 adding, 61, 62
 in division, 65–66
 multiplying, 63–64, 75–77
 place-value understanding for, 55–60, 67–68
 subtracting, 61, 62
Multiples, in Grade 4 activity, 37–39
Multiplication
 area model method for, 63–64
 arrays in, 16, 17, 19, 24, 85
 for calculating volume, 192–93, 195, 196
 grade 3
  in measurement and data domain, 148, 149, 152–53,
   154–55
  in number and operations in base ten domain, 54
  in operations and algebraic thinking domain, 16–17,
   19–20, 21–22, 23–24, 25–26, 27–28, 29, 30–31
 grade 4
  in number and operations—fractions domain, 107–9,
   110–11
  in number and operations in base ten domain, 55–56,
   63–64
  in operations and algebraic thinking domain, 32, 33,
   34–35
 grade 5
  in number and operations—fractions domain, 122–23,
   124–26, 127–28, 129–30, 131–32

Multiplication, grade 5 (continued)
    in number and operations in base ten domain, 68–69,
       75–77
  properties of, 23–24
Multiplication facts, 154
Multiplication grid, for pattern practice, 30–31
Multiplication Masterpieces, 16–17
Multiply—How & Why, 34–35
Multiplying Menace, the Revenge of Rumpelstiltskin, 109
Multiply with Pattern Blocks, 107–8
Mummies and Math, 194–96
Mummy Math: An Adventure in Geometry, 194–96
My Friends & Me, 82–83
My Whole Hexagon, 90–91

N

No Problem!, 75–77
Number and operations—fractions, 81–138
  Grade 3, 81, 82–94
    Fraction Fun, 93–94
    Jump Rope Fractions, 87–88
    Mr. Klee's Squares, 85–86
    My Friends & Me, 82–83
    My Whole Hexagon, 90–91
    Paper-Folded Fractions, 83–85
    Paper Plate Fractions, 88–90
    A Ribbon Ruler Tool, 86–87
    Stacking Wholes, 92–93
  Grade 4, 81, 95–116
    Another Serving of Cake, Please, 98–99
    Fishy Bowls, 112
    Fraction Rulers, 102–3
    Fraction Stories & Rulers, 105–7
    Fraction Teams, 97
    Full House, 110–11
    How Does Your Garden Grow?, 115–16
    It's a Piece of Cake, 95–96
    Let the Games Begin!, 104–5
    Like Denominator Addition, 99–100
    Loops & Loops, 113
    Multiply with Pattern Blocks, 107–8
    Pattern Blocks, Again!, 108–9
    Powerful Kid-Created Problems, 111
    Roll to a Meter, 114–15
    What's the Whole?, 101
  Grade 5, 81, 117–38
    Building Fractions with Box Turtles, 117–19
    Doorbell Rang, 120–21
    The Lion's Share, 133–34, *221*
    Mr. Franklin's Fractions, 135–36
    Old-Fashioned Dessert Time, 137–38
    Painting Turtles, 131–32
    Paper Plates Show the Way, 127–28
    Reasoning with Fractions, 129–30
    Thinking About Groups, 122–23
    Tiles to the Rescue, 124–26

Unlike Denominator Addition, 119–20
Number and operations in base ten, 49–80
  Grade 3, 50–54
    The Compensation Strategy, 52–53
    Hundred Charts, 54
    Race Car Rounding, 50–51
  Grade 4, 55–66
    Area Model Method—Multiplication, 63–64
    Equal Shmequal, 57–58
    Place-Value Rounding, 59–60
    Puzzle It Out, 61–62
    A Remainder of One, 65–66
    Ten Times the Money, 55–56
    Think & Fill in the Blank, 62
  Grade 5, 67–80
    Bring Out the Coins, 67–68
    Go Figure, Father Time, 77–78
    If America Were a Village, 72–73
    May I Take Your Order?, 79–80
    A Milliliter Drop, 70–71, *220*
    No Problem!, 75–77
    To the Point, 68–69
    Rounding Decimals, 74–75
Number facts
  multiplication, 154
  subtraction, 53
Number lines, 9
  fractional
    adding-machine tape for, 97
    jump rope as, 87–88, 89
    ribbon for, 86–87
  for modeling division, 20
  for modeling subtraction, 52
Number names, 57, 58, 71
Numerators
  adding, 113, 120
  explaining, 82, 129, 130
  in unit fractions, 99, 100
Numerical expressions, writing and interpreting, in Grade 5,
  42–46
Nursery rhymes, 154–55

O

Obtuse angles, 139, 173, 204, 215
Old-Fashioned Dessert Time, 137–38
One Is a Snail, Ten Is a Crab, 42–43
Operations and algebraic thinking, 15–48
  Grade 3, 15, 16–31
    Amazing Arrays, 17
    Centipede's 100 Shoes, 21–22
    Cheetah Math, 27–28
    Divvy Up!, 18–19
    Dreams of Division, 25–26
    Martha Sells Letters for $100, 28–29
    Mix It Up!, 23–24
    Multiplication Masterpieces, 16–17

Operations and algebraic thinking *(continued)*
    Practicing with Patterns, 30–31
    Six-Dinner Sid, 19–20
  Grade 4, 15, 32–41
    Add Comparison to Your Skills, 32–33
    More Than Two Ways, 37–39
    Multiply—How & Why, 34–35
    Problems with Seeds, 36
    Toothpick Triangles, 40–41
  Grade 5, 15, 42–48
    Brackets, 44
    Counting Feet, 42–43
    Decomposing Numbers, 44–45
    Express Yourself, 45–46
    Patterns with Rules, 47–48, 213
Ordered pairs, 212, 213
Origin, on coordinate grid, 210, 212

P

Pablo Picasso: Breaking All the Rules, 176–77
Painting activities
  Mr. Klee's Squares, 85–86
  Multiplication Masterpieces, 16–17
Painting Turtles, 131–32
Paper-Folded Fractions, 83–85
Paper Plate Fractions, 88–90
Paper Plates Show the Way, 127–28
Paper Quilt, 202–3
Parallel lines, in geometry domain, 204
Parallelograms, 200, 201, 202, 215
Parentheses, 42–43, 44, 45–46, 212
Pastry School in Paris: An Adventure in Capacity, 167
Pattern blocks, 9
  angle measurement and, 172–73, 178–79
  for coordinate grid activity, 210, 211
  for decomposing fractions, 101
  for decomposing numbers, 122–23
  for dividing fractions, 137, 138
  for exploring equivalent fractions, 90–91, 92–93
  for multiplying fraction by whole number, 107–9
Pattern Blocks, Again!, 108–9
Pattern Blocks, Multiply with, 107–8
Patterns, in operations and algebraic thinking domain, 15, 30–31, 40–41, 47–48
Patterns, Practicing with, 30–31
Patterns with Rules, 47–48, 213
Paul Klee (Getting to Know the World's Greatest Artists), 16, 85–86
Peacock Power!, 206–8, *224*
Pentominoes, 197–98
Pentomino Volume, for Grade 5 assessment, 230, *230*
Perimeter, 159–61, 162–63, 168–69, 195–96, 197–98
Perpendicular lines, 204, 207
Piaget, Jean, 8
Pictorial stage of reasoning, 3, 7–8, 8, 15
Picture Pie, 117–19

Pizza by the Slice, for Grade 3 assessment, *225,* 226
Pizza Problem, The, for Grade 5 assessment, *229,* 230
Place value
  grade 3 activities for understanding
    The Compensation Strategy, 52–53
    Hundred Charts, 54
    Race Car Rounding, 50–51
  grade 4 activities for understanding
    Area Model Method—Multiplication, 63–64
    Equal Shmequal, 57–58
    Place-Value Rounding, 59–60
    Puzzle It Out, 61–62
    A Remainder of One, 65–66
    Ten Times the Money, 55–56
    Think & Fill in the Blank, 62
  grade 5 activities for understanding
    Bring Out the Coins, 67–68
    Go Figure, Father Time, 77–78
    If America Were a Village, 72–73
    May I Take Your Order?, 79–80
    A Milliliter Drop, 70–71
    No Problem!, 75–77
    To the Point, 68–69
    Rounding Decimals, 74–75
Place-value chart, 57, 58, 67–68, 70, *220*
Place-value materials, 9
Place-Value Rounding, 59–60
Place-value strips, 59–60
Plates, Angles in, 174–75
Points, plotting, 212
Polygons, 201, 216–17
Powerful Kid-Created Problems, 111
Practicing with Patterns, 30–31
Prisms, 188, 189, 189, 190, 192, 193, 194–96
Problems with Seeds, 36
Protractor Measurement, 178–79
Protractors, 9, 174, 178–79, 179
Pulling apart numbers. *See* Decomposing numbers
Puzzle It Out, 61–62

Q

Quadrilaterals, 200–201, 215
Questions, taking time for, 9–10
Quilt, Hawaiian, 208–9
Quilt, Paper, 202–3

R

Race Car Rounding, 50–51
Ray, in geometry domain, 204
Real-world applications
  of geometry, 199
  of measurement and data, 139
Reasoning with Fractions, 129–30
Recipes, dividing ingredients in, 137–38
Rectangles
  area of, 124–26

Rectangles (continued)
    attributes of, 202
    fractions and, 81, 99–100, 119–20, 129–30
    in paper-folding activity, 203
    plotting, on coordinate grid, 212
Rectilinear Place to Play, A, 158–59
Remainder of One, A (activity), 65–66
Remainder of One, A (book), 65–66
Remainders, division with, 65–66
Renzulli, Joseph, 12
Rhombus, attributes of, 201, 202
Ribbon Ruler Tool, A, 86–87
Roll to a Meter, 114–15
Rounding, Place-Value, 59–60
Rounding, Race Car, 50–51
Rounding Decimals, 74–75
Rounding numbers
    decimals, 74–75
    to nearest tens, 50–51
    place-value strips for, 59–60
    Quick Tip for, 51
Ruler, 9
    angle, 9, 174, 176, 178
    as fraction reference tool, 87
Rulers, Fraction, 102–3, 104, 105, 106, 106, 137, 138
Rulers, Fraction Stories &, 105–7
Rules, Patterns with, 47–48
Rumford Complete Cook Book, The, 137

S

Seeds, Problems with, 36
Shape patterns, geometric, building, 40–41
Shapes. *See also specific shapes*
    angles as, 172–73
    attributes of, 200–201, 202
    for exploring fractions, 90–91, 92–93
    fractions as part of, 81
    perimeter of, 159–63
    rectilinear, 158–59
Show It Two Ways, for Grade 4 assessment, *227,* 228
Singapore, 8
Six-Dinner Sid (activity), 19–20
Six-Dinner Sid (book), 19–20
Skip counting, on hundred chart, 54
Small, Marian, 10
Smithsonian Handbooks: Butterflies and Moths, 169–71
Snap/interlocking cubes, for making pentomino shapes, 197, 198
Sort It Out, 214–15
Sparkles' Mix-Up Game Cards, 166, 167, *223*
Sparkles' Mix-Up Mess!, 166–67, *223*
Square
    attributes of, 201, 202
    fractions as part of, 81
Stacking Wholes, 92–93

Standards for Mathematical Practice
    identified for activities, 3, 7
    list of, 7
    purpose of, 6–7
Stepping onto Patterns, for Grade 4 assessment, *227,* 228
Storybook Character Problems, 168–69
Storytelling, for teaching fractions, 96
Story writing, about fractions multiplied by whole number, 111
Student engagement in math, importance of, 7
Subtraction
    of angles, 180, 181
    grade 3, in number and operations in base ten domain, 52–53
    grade 4
        in number and operations—fractions domain, 98–99, 105, 106
        in number and operations in base ten domain, 61, 62
    within 1000, compensation strategy for, 52–53
Sweet Clara and the Freedom Quilt, 202–3
Symmetry, 208–9
Synonyms, in math, 189

T

Talking mathematically, 12, 13. *See also* Vocabulary terms
T charts, 40, 41, 47, 48
Ten Minute Cake recipe, 138
Ten Times the Money, 55–56
There's a Formula for That!, 152–53
Think & Fill in the Blank, 62
Thinking About Groups, 122–23
Tiles, 9
    for area measurement, 124–26, 148, 149, 152–53, 156, 157
    for division problem, 65
    for making pentomino shapes, 197
    for modeling division, 21, 22
    for modeling multiplication, 21, 22
Tiles to the Rescue, 124–26
Time activities
    Get Up and Go!, 140–41
    Go Figure, Father Time, 77–78
Time Is Passing Data Sheet, 140, 141, *222*
TIMSS (Trends in International Mathematics and Science) study, 8
Tools. *See also specific tools*
    fraction reference, 86–87, 87
    managing use of, 9
    needed in math journal, 30
Toothpicks
    for building shape patterns, 40–41
    for rounding activity, 74, 75
Toothpick Triangles, 40–41
To the Point, 68–69
Trapezoid, attributes of, 199, 200, 201, 211, 215
Trends in International Mathematics and Science (TIMSS) study, 8

Triangles
    building patterns with, 40–41
    categories and sub-categories of, 214–15
    in geometry domain, 199, 203, 204, 205, 206–7
    reviewing attributes of, 214
Trip to the Park, A, for Grade 3 assessment, *225,* 226
Turtles, Box, Building Fractions with, 117–19
Turtles, Painting, 131–32
Two Ways to Count to Ten, a Liberian Folk Tale, 37–39

U

Understanding math, keys to, 7, 9–10, 11–12
Unifix cubes
    for demonstrating division, 120–21
    for modeling associative property, 23–24
    for multiplying decimal by whole number, 68–69
Unknown group size, methods of finding, 32–33, 34, 35
Unknown number of groups, methods of finding, 33, 34, 35
Unknown products, methods of finding, 32, 34, 35
Unlike Denominator Addition, 119–20

V

Variations, purpose of, 4
Vincent Van Gogh: Sunflowers and Swirly Stars (Smart About Art),
    176–77
Vocabulary terms, <u>13</u>, <u>24</u>, <u>189</u>, <u>215</u>
Volume, 142, 143, 186–98

W

Weight, 57–58
What Is a Square?, for Grade 5 assessment, 230, *230*
What's the Whole?, 101
Whole numbers
    dividing, by unit fractions, 133–34, 135–36, 137–38
    expressed as fractions, 92–93
    multi-digit
        adding, 61, 62
        in division, 65–66
        multiplying, 63–64, 75–77
        place-value understanding for, 55–60, 67–68
        subtracting, 61, 62
    multiplied by a fraction, 107–11, 124–25, 127–28, 129–30
    multiplying decimal by, 68–69
Wing It!, 169–71
Wright 3, The, 197–98
Write About It activities, purpose of, 4
Writing mathematically, 12, 13. *See also* Math journals

X

x-axis, on coordinate grid, 47, 48, 210, 211, 212

Y

y-axis, on coordinate grid, 48, 210, 211, 212

SDE

Professional Development

Who Is SDE?

Staff Development for Educators is America's leading provider of professional development for PreK through Grade 12 educators. We believe that educators have the most important job in the world. That's why we're dedicated to empowering educators with sustained PD that is not only research based, innovative, and rigorous, but also practical, motivating, and fun.

We offer:

- Expertise in the most relevant cutting-edge topics and global trends facing educators today

- Access to over 300 engaging and inspiring educational experts

- A variety of PD options to fit how you learn best and that match your budget and schedule.

Educators need flexibility and variety in accessing professional development.

That's why SDE provides multiple formats to fit how you learn best.

PD Events

- Single-Topic Workshops
- Multi-Topic Workshops
- National and Regional Conferences
- Unconference Facilitation
- In-Depth Institutes
- Train-the-Trainer Institutes

Onsite PD

- Single-Topic Workshops
- Multi-Topic Workshops
- Customized Conferences
- In-Depth Institutes
- Train-the-Trainer Institutes
- Co-Teaching
- Professional Learning Communities
- Modeling/Observation
- Job Embedded Coaching

PD Resources

- Books
- Digital Resources & e-Books
- Manipulatives
- Apps
- Games

Web-Based PD

- Webinars
- Online Courses
- Flipped Workshops
- Blended Learning

Expect **Extra-ordinary**

Together let's create extraordinary classrooms.

SDE **Staff Development** *for* **EDUCATORS**™

Serving the professional development needs of extraordinary educators.

1-877-388-2054 | www.SDE.com